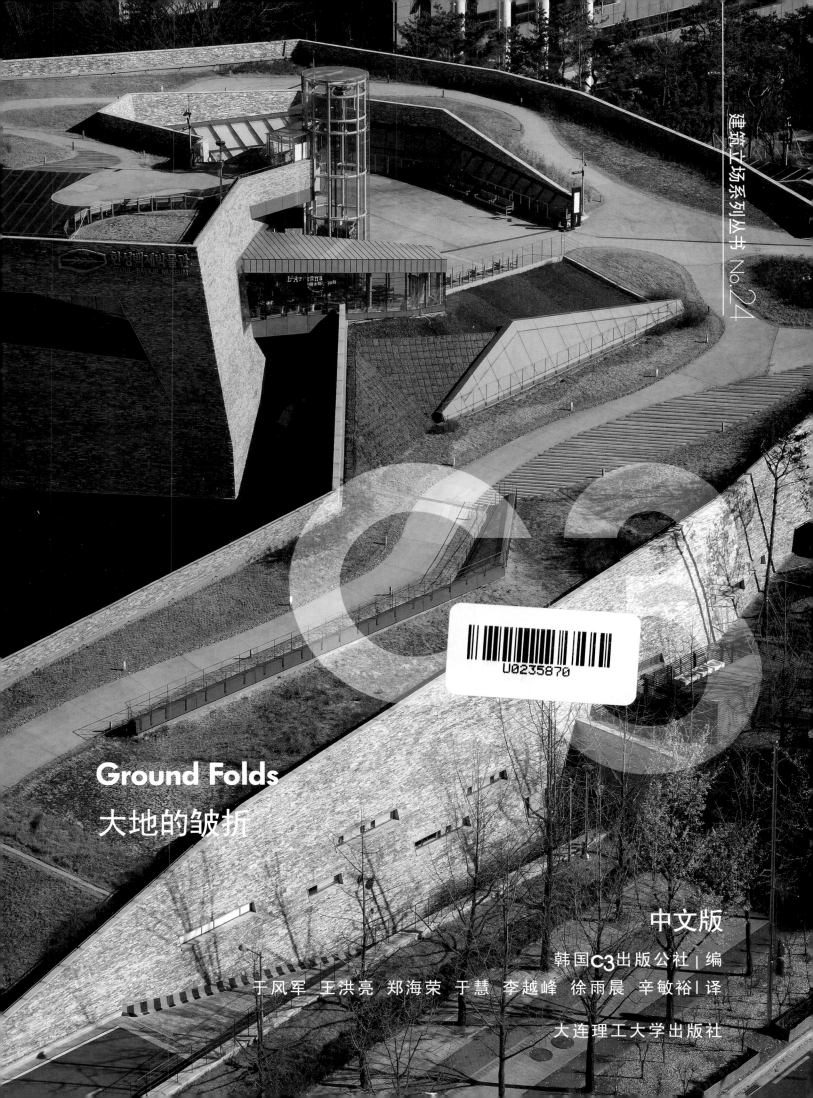

Ground Folds
大地的皱折

建筑立场系列丛书 No.24

中文版

韩国C3出版公社 | 编

于风军 王洪亮 郑海荣 于慧 李越峰 徐雨晨 辛敏裕 | 译

大连理工大学出版社

4 资讯
004 智能高速公路 _ Studio Roosegaarde
005 场景感应器 _ James Murray + Shota Vashakmadze
008 沃勒小溪 _ Michael Van Valkenburgh Associates

12 打破墨守成规
012 打破墨守成规的居住方式 _ Silvio Carta
018 潜水员的玻璃屋 _ Naf Architect & Design
030 瓷玩偶工作室 _ UID Architects
042 户田屋 _ Office of Kimihiko Okada
052 Tenjinyama工作室 _ Ikimono Architects

64 大地的皱折
064 在大地的皱折里 _ Marco Atzori
070 首尔百济博物馆 _ G.S Architects & Associates
086 Monteagudo博物馆 _ Amann-Cánovas-Maruri
100 布鲁克林植物园游客中心 _ Weiss/Manfredi
114 VanDusen植物园游客中心 _ Perkins + Will
128 象征性纪念物 _ TEN Arquitectos

138 都市改造与协调发展
138 解码本土文化：相互调和的建筑设计 _ Nelson Mota
144 Buso公寓 _ dmvA Architecten
156 模式化住宅 _ IND
168 青浦青少年活动中心 _ Atelier Deshaus

182 建筑师索引

News
004 Smart Highway _ Studio Roosegaarde
005 Scene Sensor _ James Murray + Shota Vashakmadze
008 Waller Creek _ Michael Van Valkenburgh Associates

Breaking the Stereotype
012 *Breaking Stereotypes of Living _ Silvio Carta*
018 Glass House for Diver _ Naf Architect & Design
030 Atelier-Bisque Doll _ UID Architects
042 Toda House _ Office of Kimihiko Okada
052 Atelier Tenjinyama _ Ikimono Architects

Ground Folds
064 *In the Folds of the Ground _ Marco Atzori*
070 HanSeong BaekJe Museum _ G.S Architects & Associates
086 Monteagudo Museum _ Amann - Cánovas - Maruri
100 Brooklyn Botanic Garden Visitor Center _ Weiss/Manfredi
114 VanDusen Botanic Garden Visitor Center _ Perkins + Will
128 Emblematic Monument _ TEN Arquitectos

Architecture and Recipro-City
138 *Decoding the Vernacular: An Architecture of Reciprocity _ Nelson Mota*
144 Apartments Buso _ dmvA Architecten
156 Pattern Housing _ IND
168 Qingpu Youth Center _ Atelier Deshaus

182 Index

未来概念 FUTURE CONCEPT

智能高速公路 _Studio Roosegaarde

设计师Daan Roosegaarde喜欢创新，他和Heijmans Infrastructure公司在"荷兰设计周"上展示了首批"智能高速公路"原型，并且获得了"荷兰设计大奖最佳未来概念奖"。这些智能高速公路将于2013年年中在荷兰全部建成。现在，他们正使用最新技术在欧洲修建第一条"智能高速公路"，这条公路更具有可持续性，也更为安全，更加直观化。

在未来的5年中，一些创新设计，如"黑暗中发光的公路""动态涂料""交互灯""感应优先车道"和"风力灯"等等，都将成为现实。这些创新设计目的在于通过使用符合具体交通情况的交互灯、智能能源和道路标志，使道路拥有更好的可持续性和互动性。

"黑暗中发光的公路"装置的路径经过一种特殊的张紧振子强制振动发光源粉末处理，无需额外照明就能在黑暗中发光。发光粉在白天接受阳光的照射后，在夜间发光可长达10小时，使夜间道路的轮廓清晰可见。"动态涂料"所使用的涂料会随着温度的变化在路面呈现不同的提示图案，直接向驾驶员提供相关而又充分的交通信息。例如，如果天气寒冷，道路湿滑，路面就会出现冰晶图案。

虽然"动态涂料"和"黑暗中发光的公路"将在明年实现，但是在荷兰南部城市埃因霍温举行的"荷兰设计周"上，"智能高速公路"的雏形已经面向公众开放。

Roosegaarde工作室和Heijmans Infrastructure公司的合作为创新产业和那些渴望创新的跨国公司树立了真正的典范。Roosegaarde工作室的设计和互动以及Heijmans Infrastructure公司的专业知识和技术可谓珠联璧合。尽管两者有天壤之别，但是他们所关注的是过程，而不是产品，而这个过程就是革新荷兰的景观。

Smart Highway

Designer and innovator Daan Roosegaarde and Heijmans Infrastructure presented the first prototypes of the "Smart Highway" during the Dutch Design Week. Using the latest techniques, they are building the first "Smart Highway" in Europe, roads that are more sustainable, safe and intuitive. Selected "Best Future Concept" by the Dutch Design Awards, these highways will be realized mid 2013 in the Netherlands.

Innovative designs such as the Glow-in-the-Dark Road, Dynamic Paint, Interactive Light, Induction Priority Lane and the Wind Light will be realized within the following five years. The goal is to make roads more sustainable and interactive by using interactive lights, smart energy and road signs that adapt to specific traffic situations.

The pathways of the Glow-in-the-Dark Roads are treated with a special fotoluminising powder making extra lighting unnecessary. Charged in the day light, the Glow-in-the-Dark Road illuminates the contours of the road at night up to 10 hours. Dynamic Paint, paint that becomes visible in response to temperature fluctuations, enables the surface of the roads to communicate relevant and adequate traffic information directly to the drivers. For example, ice-crystals become visible on the surface of the road when it is cold and slippery.

Although Dynamic Paint and Glow-in-the-Dark Road will be realized next year, the first pieces of the "Smart Highway" are open for the public during the Dutch Design Week in Eindhoven, the Netherlands. The unique collaboration between Roosegaarde and Heijmans is a true example for the creative industry and those multinationals with a desire for innovation. The design and interactivity from Studio Roosegaarde and the specific knowledge and craftsmanship of Heijmans combine the best of the worlds. Despite their big differences, they focus on the process instead of the product and that is innovating the Dutch landscape.

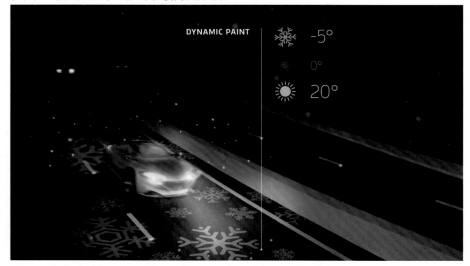

生态能源 ECOLOGICAL ENERGY

场景感应器 _James Murray+Shota Vashakmadze

在纽约Freshkills公园的水面上下，人与生态能源的关键性互动催生了复杂的"环境流"，使我们不得不提出如下问题，即怎样感应、引导和驾驭生态能源，揭示生态能源相互之间有益的波动。

场景感应器位于两股环境流的交叉点，既连接了两处完全不一样的地貌，同时又将其隔离开来：作为通道屏风，场景感应器驾驭风流穿过潮汐动脉；作为一个优势位置，场景感应器呈现了穿梭往返于公园的行人流。这两个作用合二为一，如同镜子与窗的组合，向Freshkills公园自身反射并展示了其起伏的风光景色。

南北土堤之间形成的生态流通道，是最佳的风流走廊。建筑师设计了一个风映射屏幕来检测和控制利用这一自然通道的能源。两块平板分别与两个土堤喇叭形的基座对齐排列，横跨这个垃圾填埋场的山帽之间，与塔的高度一样，耸立于吃水线之上，从而不会破坏当地生态系统。

通道屏幕组成了嵌板的框架，嵌板可以随风自由弯曲，独立做出反应，同时，作为一处场地，可以展示出更大规模的风流。屏幕像素成为揭示风的变幻莫测和反映能量收集波动的指数。每面反光金属网嵌板都与压电线相互交织，能把动能转换成电流。游客垂直于通道屏幕来进行移动，穿过公园唯一一座桥时所做的机械力通过压电换能器来捕捉，并把行人流嵌入在生态场景中。

在一个春日，通过这些交互过程所发的电足以满足1200户人家的用电需求。

当夜幕降临，通道屏幕之间由灯组成的网格将取代日光的反射，显示出白天所生成的能量。现有的桥成为不断被照亮的风流经过的有利位置，而场景感应器则提供了分布在网格结构内部边界的情景片段。这些透视点通过光线网格来切开路径，把风流的覆盖物横跨其上，它们交织在一起，使这片景观极具意义。

场景感应器的众多功能作用，如在透明和不透明间转换、在明暗之间转换、在反射和折射间转换，使其具有镜子与窗口的双重身份。作为镜子，它可以监视能量收集方面的振幅，并把振幅投影到周围景色的倒影上。然而，作为窗口，它勾勒出原本看不见的景色，把短暂的纹理有形化，为欣赏Freshkills公园的优美景色提供了前所未有的角度。

Scene Sensor

Key interactions of human and ecological energies, above and below the surface of Freshkills, drive complex environmental flows, allowing us to question how to sense, channel, and harness their energies, revealing their interconnected fluctuations in beneficial ways.

Scene Sensor situates itself at the intersection of flows joining and separating opposing landforms: as a channel screen, harnessing the flows of wind through the tidal artery, and as vantage points, staging crosswise pedestrian flows through the park, the two acting in combination as a mirror-window, reflecting and revealing the scene of Freshkills' fluctuating landscape back to itself.

Between the north and east mounds a channel of ecological flows defines a corridor of optimum wind flow. To sense and harness the energies inherent to this conduit we propose a wind mapping screen. Two planes, aligned to each mound's flare

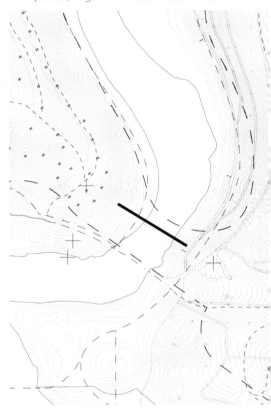

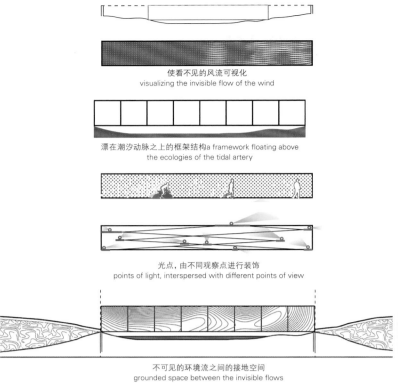

使看不见的风流可视化
visualizing the invisible flow of the wind

漂在潮汐动脉之上的框架结构 a framework floating above the ecologies of the tidal artery

光点，由不同观察点进行装饰
points of light, interspersed with different points of view

不可见的环境流之间的接地空间
grounded space between the invisible flows

项目名称：Scene Sensor
地点：Atlanta, USA
建筑师：James Murray, Shota Vashakmadze
用地面积：400,000m²
建筑面积：1,800m²
总楼面面积：3,000m²
设计时间：2012

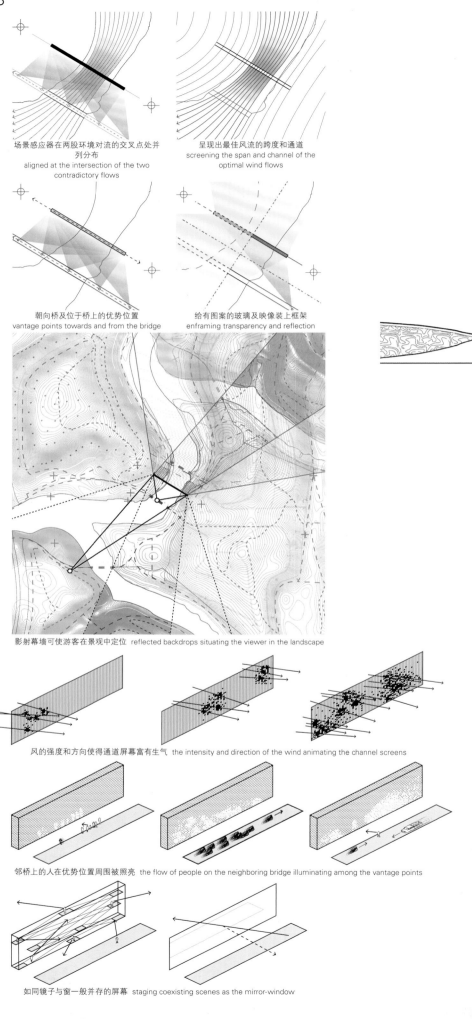

场景感应器在两股环境对流的交叉点处并列分布
aligned at the intersection of the two contradictory flows

呈现出最佳风流的跨度和通道
screening the span and channel of the optimal wind flows

朝向桥及位于桥上的优势位置
vantage points towards and from the bridge

给有图案的玻璃及映像装上框架
enframing transparency and reflection

影射幕墙可使游客在景观中定位 reflected backdrops situating the viewer in the landscape

风的强度和方向使得通道屏幕富有生气 the intensity and direction of the wind animating the channel screens

邻桥上的人在优势位置周围被照亮 the flow of people on the neighboring bridge illuminating among the vantage points

如同镜子与窗一般并存的屏幕 staging coexisting scenes as the mirror-window

station, span the width between the landfill caps and reach the height of the towers, lifting above the waterline without disturbing the local ecosystem.

The channel screens compose a framework for panels free to bend with the wind, reacting individually while articulating larger scale flows as a field. The pixels become an index that reveals the shifting winds and maps fluctuations in the resolute collection of energy. Each panel of reflective metallic mesh, interwoven with piezoelectric wires, transforms forces of motion into electrical current. Perpendicular to the channel screens, the movement of visitors across the park's only bridge presents a mechanical force to be harvested through piezoelectric transducers, embedding pedestrian flows within the ecological scene. On a spring day, the energy collected through these intersecting processes would be enough to power twelve hundred households.

As evening sets in, a mesh of lights between the channel screens replace reflections of the daylight, displaying a memory of the generated energy. The existing bridge now operates as a vantage point for the constantly illuminated wind flow while Scene Sensor affords discrete views ramping within and along the meshwork. These perspectival points carve paths through the mesh of light, situating an overlay of flows above and across, woven together to make sense of the landscape.

Scene Sensor's impositions of shifting transparency and opacity, light and darkness, and reflection and refraction stage a doubled identity of the mirror-window. As a mirror, it monitors oscillations in energy collection and projects them onto a reflection of surrounding sights. Yet as a window, it frames the unseen, materializing ephemeral textures while affording unprecedented apertures into Freshkills' scenery.

城市重生 URBAN REGENERATION

沃勒小溪 _Michael Van Valkenburgh Associates

沃勒小溪是一条狭窄的城市河岸走廊，蜿蜒曲折2414m，流经奥斯汀市中心。多年来，小溪本身饱受侵蚀、外来物种入侵和洪水之苦，从物理上和文化上与周围的城市分隔开来。2011年，一条新的隧道开始在小溪下方建造，这会有效地将113 312m²的城市用地从泛滥平原上移走，保持常年水流不断，并且防止进一步侵蚀。

获胜的提案把经过工程改造后、修建了隧道的小溪看成是奥斯汀丰富城市生活不可或缺的生态系统。设计理念是将全新的沃勒小溪进行扩建，使其延伸至公园链（嵌入五个相连的区域：格子区域、小树林、隘路、庇护所和汇合点）中。

格子区域

六座轻质易安装的人行桥组成格子结构，横跨沃勒小溪入口，形成了公园链的最南端。格子结构位于湖和小溪之间的狭小空间内，是市中心、奥斯汀东部以及人行道和自行车道之间至关重要的绿色纽带，为城市居民提供了一套全新的散步、跑步、骑自行车、社交和通勤的生活方式。

小树林

如果说呈线性连为一体的格子结构鼓励人们前前后后、上上下下地运动的话，那么小树林就是一个露天"房间"，为人们提供了一处停下脚步的小憩之所。小树林里的橡树郁郁葱葱，充满活力，从街面到小溪种满了整个斜坡。人们在这儿举行各种各样、精彩纷呈的社区活动，包括电影之夜、露天市场和户外展览等，这些活动有的是有组织有计划的，也有自发性的。

隘路

隘路位于整个公园链的中心，是沿沃勒小溪两岸新开发的区域，这一新区域的开发得益于隧道工程消除了洪水的威胁。以前住宅的阳台如沸水锅一般，酷热难耐，现在，住在面向小溪的住宅里的住户可以利用这一新的机遇，到户外坐坐、在小溪两岸现存的建筑中建造零售和艺术空间。

庇护所

栖息地庇护所成为独具特色的典范，告诉人们新的城市基础设施可以保持而不是削弱它所在的自然环境的质量，并且向人们呈现发现和探索的独特机会。桥的码头被设计建造成一台"生态机器"，用来处理来自于路面和相邻发展区域的雨水，然后再把雨水排入湿地栖息地。湿地栖息地位于一座雕塑式的建筑内，里面可以举办许多当地教育项目开展的小溪栖息地探索活动。

汇合点

公园链中最后一环是汇合点，既是对沃勒小溪隧道的呼应，也使沃勒小溪隧道以一种完全不同的方式与周围景观融为一体。隧道引水渠的混凝土叶片用作漂浮在水面上的新草坪的支承结构，如果不是这样重建这片公共区域，它就会被隧道工程吞噬了。

连接公园链每个环节的是小溪两岸本身，小溪两岸早已经留下了奥斯汀城市建筑的有形历史。耐用的生物工程技术将应用于小溪两岸的改造，过度生长的外来入侵物种将被消除、当地原生植物将会重生，沃勒小溪将以全新面貌迎接城市居民。

项目名称：Waller Creek
地点：Austin, Texas, USA
总景观设计师：Michael Van Valkenburgh Associates
总建筑设计师：Thomas Phifer & Partners
景观建筑师：dwg.
城市设计及规划师：Greenberg Consultants
桥梁设计及结构工程师：HNTB
经济开发师：Development Strategies
当地土地使用及分区：Metcalfe, Wolff, Stuart & Williams
公共空间管理：ETM Associates LLC
艺术家：Oscar Tuazon
设计时间：2012

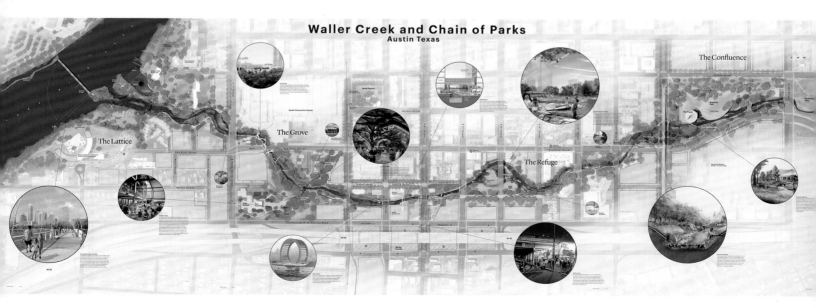

Waller Creek

Waller Creek is a narrow urban riparian corridor that meanders for 1.5 miles through the downtown of Austin. Over the years, the creek itself has suffered from erosion, invasive species, and flash flooding, and has been physically and culturally isolated from the city around it. In 2011, construction began on a new tunnel beneath the creek that will effectively remove 28 acres of the city from the floodplain, maintain constant water flow, and prevent further erosion.

The winning proposal seeks to understand the engineered post-tunnel creek as an ecological system that can be essential to the great city life in Austin. The design concept expands a renewed Waller Creek into the Chain of Parks embedded in five connected districts: the Lattice, the Grove, the Narrows, the Refuge, and the Confluence.

The Lattice

A lattice of six lightweight and easily deployable trail bridges spans the mouth of Waller Creek and forms the southernmost link of the chain. Situated at the liminal space between the lake and creek, the lattice becomes a vital green link between downtown, east Austin, and the hike and

旧剖面图1 section_old 1

旧剖面图2 section_old 2

新剖面图 section_new

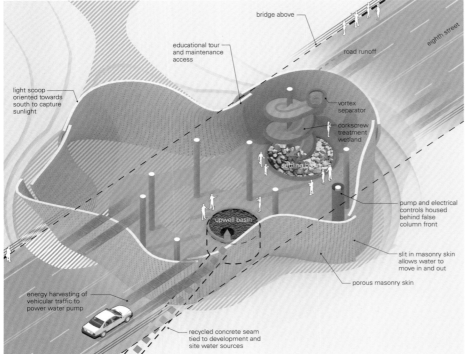

bike trail, and inspires an entirely new set of rituals for walking, running, biking, socializing and commuting.

The Grove
While the linearly connected pieces of the lattice encourage movement forwards, backwards and above, the grove offers a place for pause in the form of an open-air "room". A new grove of the living oaks fills a broad slope descending from street level to the creek. The grove hosts a tremendous variety of community programs, both planned and spontaneous, including movie nights, open-air markets and outdoor exhibitions.

The Narrows
At the center of the chain lies the narrows, an area of new development along the banks of Waller Creek, enabled by the tunnel project's elimination of the threat of flood. Following the cues of the boiling pot balcony, creek-facing properties may make use of new opportunities for outdoor seating, retail and art spaces within the existing architecture lining the creek.

The Refuge
Presenting a unique opportunity for discovery and exploration, the habitat refuge offers a distinctive model for the way in which new urban infrastructures can sustain rather than weaken the natural environment in which it sits. The bridge pier is engineered as an "ecological machine" that treats stormwater from the roadway and adjacent development before discharging it into a wetland habitat, alluringly housed in a sculptural structure that hosts explorations of creek habitat led by local educational programs.

The Confluence
Also responding to and engaging the introduction of the Waller Creek Tunnel into the landscape, but in a wholly different way, is the final link in the chain, the confluence. The concrete vane for the tunnel approach channel is used as a structural support for a new lawn that floats out over the water, reclaiming parkland that would be otherwise lost to the tunnel project.

Connecting each link of the chain are the banks of the creek itself, which have inscribed upon them the physical history of the city-building in Austin. Durable bio-engineering techniques will be employed along the banks of the creek, and the removal of overgrown invasive species and regeneration of a native plant palette will open up the creek to the city.

打破墨守成规的居住方式

Breaking Stereotypes of Living

潜水员的玻璃屋/Naf Architect & Design
瓷玩偶工作室/UID Architects
户田屋/Office of Kimihiko Okada
Tenjinyama工作室/Ikimono Architects
打破墨守成规的居住方式/Silvio Carta

Glass House for Diver/Naf Architect & Design
Atelier-Bisque Doll/UID Architects
Toda House/Office of Kimihiko Okada
Atelier Tenjinyama/Ikimono Architects
Breaking Stereotypes of Living/Silvio Carta

房屋往往都是在无形的"盒子"中设计出来的，这个无形的"盒子"反映着设计习惯、地方法律法规或设计建筑指南等。避开"常规"的结果常常会使设计项目价值倍增，并且一般来说，会得到房屋的最终使用者的积极肯定。但是什么样的房子是"普通"的房子？"普通"房子与"奢华"房子有什么区别？另外，设计师的工作，无论是从专业角度来说还是从智力水平来说，都充斥着不断的挑战：既不能对自己的所有设计想当然，也不能生搬硬套（设计、建筑结构、规范标准的）相关规定。有的人也许会说，好的指导思想就是选择两个极端的中间位置；而有的人就是否应该把设计习惯和规章制度作为建筑师解决问题的工具提出疑问，也许这个问题可以换种方式表述：如何打破限制我们"居住空间"的僵化思想？设计师往往按照固定模式或者过于简单化的居住理念来勾画普通住宅。在设计房屋之前，设计师应该去询问，并更加仔细考虑人们真正的（多种多样的）生活居住方式。人们在过去、现在的需要和对未来的憧憬所形成的范式和模型之间不断变化中寻求平衡，想要充满活力而不是一成不变地生活着。因此，如果人们随着时代的变化每天都在重新思考和重新确立新的生活方式，我们每天都生活在其中的房子为什么就应该停留在一个先入为主的想法中呢？

　　下面呈现给读者的一系列设计展示了如何处理好房屋设计的主要思想和人们对房子的固定思维模式的可能性，使读者清晰无误地了解到，设计师并没有把生活看作是固定不变的人类活动。

　　类型和模型的定义及其应用一直以来都是建筑设计行业至关重要的方面。几个世纪以来，建筑师们通过定义原型和建筑类型学，已经渐渐习惯接受空间特征、形式和用途之间的某些核心关系。

　　某些固定"形式"，或更确切地说，某些与建筑用途相关的一套套正规拘谨的特征，甚至在今天仍是设计行业的

Houses are often designed within an invisible box reflecting design habits, local regulations or handbooks' indications. To escape "normal" results often becomes an added value to the project and it is in general considered positively by the final users of the house. But what are the extents of a "regular" house project? And what does it differentiate from an "extravagant" project? Moreover, the work – both professional and intellectual – of a designer consists of a continuous challenge between the fact of not taking anything from granted about his/her design and yet following (design, constructive, normative) rules. One may argue that a good direction is a middle position between the two poles. Others can raise the question of interpretation (of design habits and rules) as the solving tool for architects. Perhaps the question can be reformulated by tackling the cliches that frame our "living spaces". The image of the common house is often conceived by following a fixed or oversimplified idea of living. Before containing space, the actual activity of living (in all its varied manners) should be questioned and more carefully considered. People conduct their lives dynamically, in a continuously changing balance between paradigms and models from the past, current needs and future ambitions. Therefore, if the act of living is being reconsidered and reestablished everyday, according to the change of times, why should the house – accommodating the living within physical boundaries – be anchored to a preconceived idea?

The presented projects display a range of possibilities of dealing with the main idea of the house and its stereotypes, offering clear reconsiderations of living as not fixed human activity.

一项重要参考。设计手册和设计指南就是显而易见的例子，它们向学徒建筑师描述了如何设计医院、博物馆和房屋等。这些"如何设计"的圣经（常常以建筑标准和地方法规来支撑）明确地解释说明了如何根据建筑的最终用途来进行设计。房间的大小和数量、空间的分配和固定元素（走廊、前厅、楼梯）之间的距离等都是书中所讲的设计应该考虑的一些基本参数。换句话说，设计应该按照手册中所讲的所有数据和指示去做。私人住宅的标准化布局就是这一现象的典型例证。私人住宅（至少）包括卧室、起居室、厨房和卫生间，外加入口、前厅和存储区等服务性空间。这种标准化设计是基于这样一种理念：建筑最终使用者对建筑的要求基本上大同小异，因此建筑物是可以预见的。虽然按图索骥的指南设计法有许多可取之处（如设计和建设所需时间相对较短、普遍可行、广泛的市场认可度和相对经济），但是人们对这一问题多加思考后，也许就会得出完全不一样的结论。

也许有人会说，设计指南书籍的多数读者没有批判性地对待书中所提供的数据和所列举的现象。从这一方面来看，我们可以看出设计师如何几乎毫无意识地把自己的设计置于一个固定的框架之内。这一框架的标签就是"优秀建筑的规则"，它列举了一整套大家普遍认为设计这类建筑应具有的建筑特征。这样的设计，甚至在其诞生之前，似乎就已经陷入了一个无形的牢笼之中，里面充满了指导方针、规则章程和设计惯例。这些对项目的管理和施工阶段确实有很大帮助，但是也大大地限制了设计创意和创新的维度。预先确定的"建筑参考"最终形成了一系列意料之中的建筑特征，但是现在看来却事先限制了项目的性能。在设计开始之前，实际上就有一个类型理念表明了设计进一步发展的方向，但是与此同时，也减少了更多可能的设计选择。单就住宅设计而言，住宅项目遵循的理念是满足一个"家"普遍具有的所有条件。从一定程度上来说，这一概念是一直以来人们脑海中根深蒂固的住宅样式的结果。大家都知道房子应该是什么样子，不需要任何解释[1]。

然而，尽管一些建筑师把指南里的类型和模型看作是设计的目标（认为如果设计达到了模型中所包含的空间、功能和正式的要求，这个设计就是好设计），认为这样的设计才保险，但是也有一些建筑师只是把那些设计指导方针看作是进行设计工作的起点。实际上，一些

The definition and application of types and models have always been crucial aspects of the architectural profession. Down through the centuries, architects have gradually settled into certain core relationships of spatial characteristics, forms, and uses by defining archetypes and architectural typologies.
Certain fixed "forms" – or better, certain sets of formal characteristics related to use – represent an important reference in the design profession even today. A clear example is found in the form of design manuals and handbooks which describe to apprentice architects how to design a hospital, a museum, or a house. In such "how-to" bibles it is clearly explained (often with support from construction standards and local regulations) how to execute design based on final use. The size and number of rooms, spatial distributions, and mutual distances between fixed elements (corridors, anterooms, staircases) are some of the basic parameters around which one is told how the design should be conceived. In other words, the design should comply with all the data and indications found in the manual. A demonstration of this phenomenon may be found in the standard layout of private houses, which includes (at minimum) sleeping rooms, living rooms, kitchen and toilet, plus such service spaces as entrances, vestibules and storage areas. This design standardization is grounded in the idea that all requirements of a building's final users are largely held in common, and are thus predictable. Although the handbook approach has much to recommend (including relatively short time required for design and construction, general feasibility, wide market acceptance and relative economy), a more speculative side of the question exists which may allow for a reversal of this point of view. It can be argued that most readers of design handbooks do not receive the data and phenomena those books provide in a critical manner. Viewing the matter from this perspective, one may see how the designer positions his or her project, almost unconsciously, within a fixed framework characterized by the "rules of good building", meaning a set of architectural features which are commonly considered to work for the type of building being designed. The project, even before its birth, appears to have already been forced into an invisible cage made of guidelines, rules and design habits which will most assuredly help the design process through its regulation and construction phases, but which will also significantly limit its creative and innovative dimensions. Predefined "design references", resulting in a set of expected architectural features, are now shown to be constraining the project's performance, ex-ante. Prior to the project's commencement, there is in fact a typological idea which shows the direction for the design's further development, but which at the same time diminishes its potential alternatives. To focus this argument on residential design, the project of a house follows the idea that we all have of a "home". To a certain extent this conception is the

建筑师在设计中使用有时被称作"发散思维"[2]的思维模式。解决问题的方法（被叫做"聚合思维"）是基于对特定问题最有效的解决方法的研究，而发散思维则是对特定问题同时探索多种可能的解决方法，因此需要重新设定特定问题的最初情况。在这种情况下，没有明显解决办法的问题会促进整个问题的重新形成。

把发散思维应用于房屋设计使人们对"居住"（相对于仅仅占据房子而言）真正意味着什么有了更广泛的思考，为关于居住的新一轮思考提供了空间，也为这一主题提供了更广泛的方法。人们呆在自己住处这一行为不再是想当然的，而是根据其多方面的含义进行了重新思考。出现的新问题是：现在人们究竟是怎样生活的？呆在住所里的时间有多少？对住处究竟有何要求？

房子不再是为承担居住活动功能而安排得井然有序的空间；它是居住者性格特征的体现（居住者的理想抱负、恐惧、对世界和社会的真知灼见），因此对房子的设计也成为建筑师不断探求建筑师职业的界线和意义的机会。这一设计方法的主导理念在于我们对房子共有的印象不再是我们居住生活的唯一可能形式。

本书中展示给读者的设计是对设计模型和建筑类型的被动接受所做出的具体回应。

潜水员的玻璃屋由一家叫作Naf Architect & Design的日本工作室设计，可以看作是重新探索房屋的通常定义的一次尝试。玻璃屋完全忽略了房屋建筑构件在预想中可能形成的结构：通常墙所起的作用现在被一堆混凝土砌块和直立玻璃平面之间的缓冲空间取代，玻璃墙架构出室内空间。屋顶构件也没有覆盖整个房子，只在屋顶主要支撑结构上铺设了部分屋顶。各种各样的人行步道和直立的玻璃平面界定了室内和室外的空间，区分出起居空间和天井，使露台成为室内外空间的中间地带，房子由一些通常不被人熟悉的元素构造（室外所使用的材料是用于港口和工业环境的典型材料）。玻璃屋不是一个封闭的空间，看起来像是镶嵌在一个由一系列随机堆放的混凝土砌块围成的空旷空间里。另外，各种房屋建筑构件组成的结构也允许房屋有一定的不确定性。在"普通的"房子中，墙的顶部与屋顶的衔接应该是连贯一致的，但是在Naf Architect & Design工作室设计的玻璃屋中，金属横梁从屋顶向前伸出，搭在混凝土砌块上。房子的外层防护通常都会

result of a consolidated image of the house which has been built throughout history. The idea of the "shape" a house should have is commonly shared and requires no explanation[1].

However, while some architects perceive the handbook typologies and models as a destination (the project is well designed when it meets the spatial, programmatic and formal requirements contained in the model) and as a safe place to land, others consider those guidelines to be merely a starting point for their work. In fact, some architects appear to use what is sometimes called "divergent thinking"[2] in their projects. While a problem-solving approach (referred to as "convergent thinking") is based on research into the most effective solution to a given problem, divergent thinking explores a range of possible solutions at the same time, thus resetting the original scenario of the given problem. In such cases, problems with no apparent solution may facilitate the reformulation of the entire problem.

Applying divergent thinking to a house project allows for the emergence of broader reflections concerning what "living" – as opposed to merely occupying a house – really means. Such an approach makes way for a new set of speculations on living, resulting in a wider approach to the topic. The human act of staying in one's own place is not taken for granted, but is reconsidered along with all its manifold implications. New questions emerge: how do people really live nowadays? How many hours do they spend in their living space? What do they really ask of their place?

The house is no longer an organized container of activities related to living; it is the expression of the persona of the inhabitant (with his/her ambitions, fear, vision of the world and society), and it has thus become an occasion for the architect to continually question boundaries and the meaning of the architect's profession. The leading idea of this design approach is that the shared image we all have of the house is not the only possible formal response to the question of how we live.

The projects presented in this issue offer a concrete reaction to the passive acceptance of design models and construction types. The *Glass House for Diver*, designed by the Japanese studio Naf Architect & Design, can be seen as an attempt to explore afresh the common definition of the house. The Glass House neglects the expected configuration of a house's constructive elements: the function normally performed by walls is here "obtained" by means of a buffer space between the piled concrete blocks and the vertical glass surfaces framing the inner spaces. The roof elements do not cover the entire area of the house, but only partially follow the main supporting structure of the ceiling. The varied pavements, along with the vertical glazed surfaces, establish what is interior and exterior, differentiating the living spaces from the patios and marking the terraces as in-between spaces. The program of the house is thus hosted within elements that are normally unfamiliar

到达边缘，但是玻璃屋那波浪起伏的屋顶板却只铺到金属梁整个跨度的中间，给人一种不完整感。这个房子的"本质"在于让人感觉房子是嵌在一堆陌生的元素中，只是偶然用来作为一个庇护所。

如果说玻璃屋打破了人们想到住宅空间时应具有的几个"条条框框"，Ikimono建筑师事务所设计的Tenjinyama工作室所采用的设计方法在这方面就可以说是颠覆性的。尽管位于广岛的项目中居住空间的安排布置从一定程度上可辨认得出来（房屋各种各样的功能位于不同的位置，分布在房子表面的四周），然而，在Tenjinyama工作室的设计中，房屋高度完全消失不见了，让人感到在这样一个完全与建筑的外立面分离开来的独特空间里，房屋特征也变得模糊不清。Tenjinyama工作室全部功能都融合在同一个独特的空间实体中。再者，建筑的内部（我们称其为被容纳者）完全独立于建筑外壳（容器）。内部的组织架构与建筑物的外壳（在"普通的设计"中两者是紧密相关的）完全分离。外立面上开口的大小、比例和位置似乎和建筑内部的活动或功能毫无关系。从严格的功能意义上来说，Tenjinyama工作室本可以更低一些，大多数活动可分布在一楼，这样，所需要的高度将会更低一些。然而，Ikimono建筑师事务所所设计的完整高度表明该工作室也许需要那样的垂直空间来创造合理的工作条件。因此，工作室的空间配置不仅是应对工作所需，也是对混凝土外壳下整体空间的创建和包装的结果。

潜水员的玻璃屋挑战了传统的居住空间处理方式，Tenjinyama工作室质疑了传统的垂直空间结构，而由Kimihiko Okada工作室设计的户田屋则打破了第三个设计规则"普通空间性"，也就是空间的分配。实际上，房子从地面层一直螺旋上升至最上面的屋顶，与生活居住有关的所有功能一应俱全地分布在这条连续向上的路径上。通常的走廊/房间这类空间的排列层次被搁置一边，各种空间都融合在一个独特的狭长区域；任何想要把这一空间性分类的尝试都是徒劳无功的。如果从普通空间分配的角度来看户田屋，人们一定会感到困惑：房子是一个包含一切应有尽有的居住生活要求的连续的走廊，还是由一系列一个挨着一个、没有走廊的房间组成的？这种新型的空间性使人们对居住空间这一话题有了更多新的思考。户田屋没有主风景或主要区域，但却处处是风景：房子与360°全景的关系既是恒定不变的也是在不

to houses (the materials employed for the exterior are typically used in ports and industrial environments). Rather than being set within an enclosed space, this house appears to be inserted into an empty space generated by a series of randomly placed stacked concrete blocks. Moreover, the configuration that the constructive elements assume allows for a certain level of indeterminacy. In an "ordinary" house the upper part of the walls should create a continuous form with the ceiling, while in the project by Naf Architect & Design, the metal beams slide forward from the roof, landing on the concrete piles. The usual finishing of the outer envelope of the house should result in an edge, but in the Glass House the undulating roofing sheets stop in the middle of the beam spans, conferring a sense of incompletion. The "nature" of this house is characterized by the sense of being nested within alien elements, occasionally used to create a shelter.

If the Glass House breaks several "ordinary rules" for thinking about the spatiality of a house, the *Atelier Tenjinyama* by Ikimono Architects brings a completely different approach to the discussion. While the Hiroshima project maintains a certain recognizable spatial organization of the living space (the various functions are hosted in different spots and distributed around the house surface), in the Atelier Tenjinyama height simply disappears, in the sense that identities are blurred in a unique space which is absolutely disconnected from its facades. All the functions of the Atelier are merged in the same unique spatial entity. Moreover, the inside of the building – let's call it the contained – is utterly independent of the outer shell – the container. The interior organization of the program and the building envelope – which would be related in an "ordinary project" – are here completely disconnected. The size, proportion and positions of the openings in the facades seem to align with no activity or program inside. In strictly functional terms, the Atelier could have been much lower, with most activities distributed on the ground floor, so that the necessary height would have been much lower. However, the full height designed by Ikimono Architects suggests that the Atelier might need that vertical space to create the correct conditions for working. The Atelier is hence not only configured as the result of the spaces needed to perform work, but by the creation and wrapping of the overall space under the concrete envelope.

Where the Glass House questions the spatial settlement of a living space and the Atelier Tenjinyama questions its vertical configuration, the *Toda House* by Office of Kimihiko Okada breaks a third rule of "ordinary spatiality", namely its distribution. In fact, the house is conceived along a spiral progression from the ground level to the uppermost roof, where the programs related to living are distributed along a continuous path. The usual corridor/room hierarchy of spaces is here left aside, with spaces being fused into a uniquely elongated development. Any attempt to categorize

1. 有意思的是，人们很可能会因为普遍认同的主要住宅形状——四方形底座外加三角形屋顶（从斜屋顶的轮廓可见）——而对这些住宅理念盲目崇拜，然而对于另一些建筑类型，比如医院和博物馆，这样的崇拜却很难产生。
2. 发散思维与聚合思维恰恰相反，它被认知主义者广泛探索并研究。对其进行的初步探讨见英国皇家艺术学会发布的、肯·罗宾森爵士创作的影音动漫《（教育）模式转型》，或者爱德华·德·博诺著的《水平思考》，Ward Lock Education，1970年。

1. It is interesting to note that it is possible to iconize the idea of the house through its primary commonly shared shape – a square with a triangle on top (the pitched roof silhouette) – while doing so for other typologies, such as hospitals or museums, would be quite difficult.
2. Divergent thinking, as opposed to convergent thinking, has been widely explored and studied by cognitivists. For a preliminary discussion see *Changing (Education) Paradigms* by Sir Ken Robinson – a video animation by the Royal Society of Arts, or *Lateral Thinking* by Edward de Bono, Ward Lock Education, 1970.

断变化的，使房子的每一处都非常奇特和珍贵。房屋内的活动或通过对一些具体位置的专门设计来体现，或对一些空间根本不作任何处理，由居住者自己决定、随意支配。房子功能和空间分配之间的关系完全被重新定义，这样一来，房子的整体外观（房子是用来居住的空间）最终反射出一种挑战房子使用的标准方式的意愿。这一设计似乎不是只建造一套固定的空间，而是勾画一个框架，随着时间的推移，框架内的功能可以得到测验、开发和改进。

与玻璃屋一样，由UID建筑师事务所设计的位于大阪箕面市的瓷玩偶工作室展示给读者的也是房屋结构元素的不连续性。设计目的似乎不是修建能满足生活活动需要的房子，而是创造一个可在其中生活的开放空间。从这一方面来说，我们可以看到这个设计中容器和被容纳者之间实际和概念上的距离。

这两者之间的关系看起来不再泾渭分明，因为后者容纳在前者留下的间隙空间里。工作室的外形根本不能反映出其内部构成，甚至没有留下任何这方面的蛛丝马迹。从里面看，由于外立面元素的作用，建筑外部给人的感觉是扭曲的。这些非关系（视觉上和实际上）促成了不确定性，从建筑体量方面来说，其多样性也增加了这种不确定性。例如，地下室层空空如也（底部体量好像漂浮在空中），两个主要建筑体量之间也空无一物（较小的建筑体量似乎漂浮在较大的建筑体量的里面和上面）。另外，工作室的空间布局反映了建筑内部功能的变化：没有了黑白表面，内部—外部之间的关系不断受到由此形成的空置空间、缝隙和间隙缺口的挑战。

上述设计的共同之处在于设计师都意图超越传统的建筑界线，从主要构成部分（墙、屋顶、表面）和彼此之间关系（比例、位置和连接），到较小的细节，如角落、边缘或台阶等各个方面质疑建筑元素。

这些挑战最终对建筑的空间布局和审美具有重大影响，然而再细想一下，他们挑战的是人们对居住空间的惯常思维、想象和体验。

this spatiality is doomed. If the Toda House is observed through the lens of ordinary distribution, confusion ensues as for whether the house is a continuous corridor containing all the required programs, or a series of rooms following one another with no corridor. This new sort of spatiality allows for a new set of speculations in the discourse of living. The *Toda House* has no main views or noble areas: the relationship with the panorama is constant and continuously changing, rendering each spot of the house peculiar and precious. Activities are suggested by the design of specific locations or left unresolved in order for the inhabitants to create their own set of spatial decisions. The relationship between program and space-assignment is completely re-invented, so that the overall appearance of the house (the space dedicated to living) eventually reflects a will to question the standard ways of using a house. The project seems not to impose a set of fixed spaces, but to draw a framework in which functions can be tested, explored and modified over time.

Similarly to the Glass House, the *Atelier-Bisque Doll*, designed by UID Architects in Minoh, Osaka, displays a discontinuity of the constructive elements of the house. The aim seems not to be to build a conclusive shape hosting the living activities, but rather to create an open configuration in which living can be hosted. In that regard, we can observe a physical and conceptual distance between the container and the contained of the house project. The two are no longer in a clearly visible relationship, since the latter is hosted within the interstitial spaces left by the former. The outside appearance of the Atelier does not reflect the internal configuration, nor does it even provide any clues, while from the inside the outside is perceived in a distorted way, because of the external facade elements. These non-relationships (visual and physical) contribute to the creation of an indeterminacy are also emphasized by various volumetric aspects, such as the absence of material at the basement level (the bottom volume seems to float in the air) or between the two main volumes (the smaller volume seems to float inside and atop the bigger one). Moreover, the spatial configuration of this Atelier reflects the shifting of the program and functions inside: the inside-outside relationship is continuously challenged by the empties, cracks and gaps created with the absence of the black and white surfaces.

The projects here presented have in common the intention of moving beyond the boundaries of usual building, calling into question a wide range of constructive elements, from the main components (walls, roof, surfaces) and their relationships (in proportion, position and junction) with the smaller details, such as corners, edges or steps. All these challenges have a substantial impact on the spatial distribution and the aesthetics of the final results, while – in a more speculative way – they challenge usual ways of thinking, picturing and experiencing our domestic spaces. *Silvio Carta*

打破墨守成规 Breaking Stereotype

潜水员的玻璃屋
Naf Architect & Design

玻璃屋是1.0m×1.0m×1.5m的大型混凝土砌块堆叠在一起形成的一个简单的构造体系，如同在一个体块上面建造另一个体块一样，形成一个波状起伏的大型结构，类似于防波堤或者消波体块。这些大型的混凝土砌块在广岛江田岛市的任何一家水泥厂都可以生产，成本非常低廉，因为剩余的水泥既可以再利用，也可以常常用作挡土墙表面或修建牡蛎养殖所需的木筏的锚。这些大型的混凝土砌块是利用剩余的混凝土生产的，只有水泥厂有混凝土订单的时候才能生产，因为生产时总会有剩余，所以就把剩余的水泥制成混凝土砌块来销售，而不是丢弃。因为混凝土砌块的生产是根据工厂的实际生产情况所决定的，所以整个玻璃屋工程的进度也受到混凝土砌块生产速度的影响。建筑的建造不是像标准化施工程序那样按照时间表来进行，而是需要等待有足够的大型混凝土砌块才能继续施工；这种"慢建筑"在现代建筑领域被认为是荒谬可笑的。

这些混凝土砌块表面都有凹槽，便于起重机吊起搬运。不仅如此，砌块摞起后这些凹槽可形成一条上下直线，在砌块间可以安放钢筋来加强房子的抗震性能。这种堆叠方式十分标准，且富于变化，错落有致，疏密合理；有的地方可观景，可改善室内的空气流通，有的地方则形成相对私密的空间。

除了上面提到的混凝土砌块结构，建筑整体结构还划分出室内空间的屋顶和围合室内的透明玻璃。混凝土砌块上面没有屋顶，阳光散落在砌块上，反射进入室内。此外，沿着砌块的周围将种植各种不同的藤本类花卉，在不久的将来，这个防波堤形状的结构成为一座开满鲜花、铺满绿色的小山。

众多堆放在一起的大型混凝土砌块作为建筑物结构似乎太庞大，其外部轮廓也绝非寻常，这座建筑也承担着控制风、光和植物的景观作用，把这些自然元素引入室内，赋予室内空间更大的自由度，使其大大超越了传统的建筑形式。

Glass House for Diver

This building is based on a simple system of stacking large concrete blocks of 1.0m x 1.0m x 1.5m like building blocks on top of one another to make a structure with big undulation like breakwater or wave-dissipating blocks. These large concrete blocks are manufactured at any cement factory in Etajima city, Hiroshima, at a very low price as excess cements are reused and often used for the surface of retaining wall or anchors of rafts for cultured oysters. These large concrete blocks are manufactured from leftover

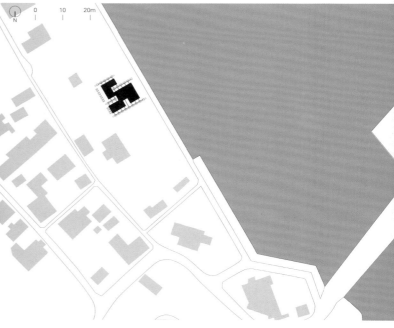

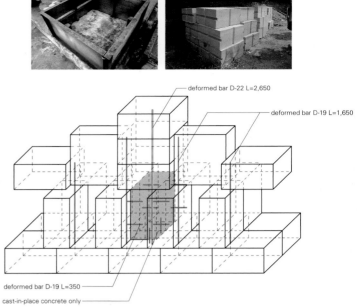

大型混凝土砌块的砖石结构
masonry construction system of the large concrete blocks

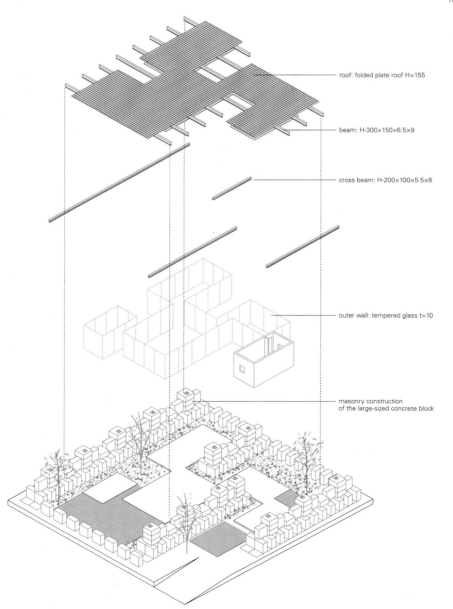

- roof: folded plate roof H=155
- beam: H-300×150×6.5×9
- cross beam: H-200×100×5.5×8
- outer wall: tempered glass t=10
- masonry construction of the large-sized concrete block

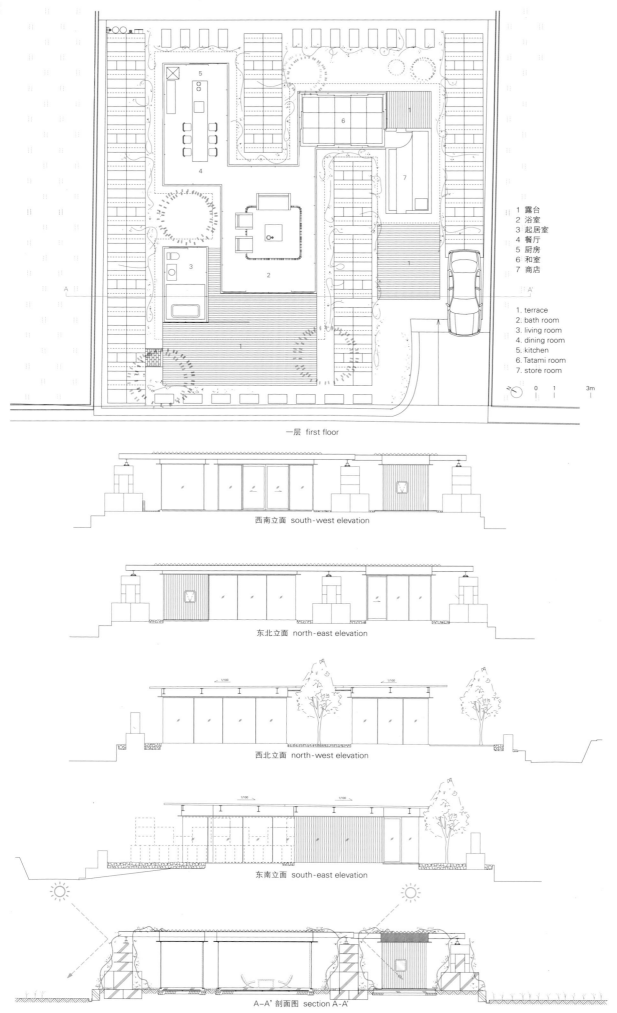

项目名称：Glass House for Diver
地点：Etajima-shi, Hiroshima, Japan
建筑师：Tetsuya Nakazono
工程师：Kenji Nawa
项目规划：villa, sanatorium
用地面积：442m²
建筑面积：104.46m²
总楼面面积：97.26m²
设计时间：2008.12—2010.4
施工时间：2010.5—2011.8
摄影师：©Toshiyuki Yano(courtesy of the architect)

concrete whenever there is an order of concrete at cement factory, as there will always be a surplus, which is sold as a product instead of discarding. As concrete blocks are manufactured at such a rate, according to the operation situation of the factories, the construction progressed according to the pace of the manufacture of the concrete blocks. The construction of this building was not based on the time schedule, as in standard construction progress, but waited for the stock of large concrete blocks to continue the work; a "slow architecture" is perceived ridiculous in modern construction work.

Groove is cut on the surface of these blocks to have them lifted by cranes. By stacking blocks to make vertical line of this groove, reinforcing steel can be placed through the blocks to secure quake resistance. This method of stacking is made standardized to give variation of directions and intervals, at a place giving view and making passage of the wind and at another place securing the privacy.

Overall composition of the building consists of roof and transparent glass to enclose interior space with the aforementioned concrete block structure. There is no roof over the concrete blocks. Therefore rays of the sun pour over the blocks, and the reflection of the light shines inside the house. Furthermore, vines of various flowers will be planted along the block, changing breakwater-like structure to a hill of flower and greenery in the future.

Mass of stacked large concrete blocks is far too enormous as a structure of a building, and the silhouette is far from ordinary. The building also has a function of landscape which controls wind, light and green, leading these factors interior and giving freedom to the interior space which exceeds the form of conventional architecture. Naf Architect & Design

瓷玩偶工作室
UID Architects

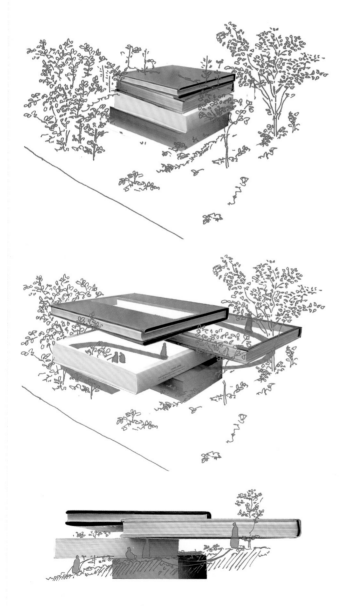

场地要求和特点

本项目既是玩偶艺术家工作室又是艺术家夫妇的居所。客户要求工作室既可用作玩偶的展廊，同时又可用作玩偶的制作空间，除此之外，它也可作为邀请朋友小聚、共度欢乐时光的空间。在居住环境方面，人们希望工作室和住处都具有开放性，同时相对于周围环境来说又保留一定程度的私密性。

建筑师还记得第一次到大阪辖区内的箕面市时，这个地方到处是绿色，不但因为当时正值春季，还因为箕面市具有严格的植物绿化标准，场地中高大、适中、低矮树木的覆盖率不低于10%，而整座城市市区建筑物的覆盖率要低于60%。

无域之疆

为了确保隐私，建筑师思考着能否通过让更多的人参与其中来实现，而不是在建筑四周修建围墙和栅栏，只能从里面向外看。

这就需要从外面创建整个空间，把邻居的绿色植物作为室外空间，而不是从里面设计必要的功能。建筑师没有选择像墙、顶壁或栅

栏之类的隔挡物，而是选择了齐腰的、不受重力影响的院墙，或者说是环绕场地的两条悬置的墙带。这种理论所建造的建筑看起来像是延伸到室外的室内空间，反对实质性的内/外区域划分。

悬置的墙带

明确地说，建筑师期望场地不用于规定内/外区域，而是由空间的多样性所决定，这种多样性来自于悬置的墙带（二层或者三层墙带堆叠在一起，围合成长方形，以保持平衡）和围墙本身。

建筑利用了场地1.2m的水平落差，两处功能区域设置在悬置墙带的重叠之处。工作室在较低区域，面北，可以看见街道；而艺术家夫妇居住在上层南部区域，并修建了一条小路，犹如一个斜坡，人们走在上面犹如走在倾斜的地面上一般。在工作室和住宅中建造一个小型盒状结构，以发挥像壁橱这样的必要功能，使这一空间看起来就像一个临街的、无边界的露天市场。三个叠放在一起的墙带把工作室和住宅融合在同一个空间里。

有方向性地重叠围墙带这个简单的设计模糊了场地的边界，使场地与邻居之间的关系具有了全新的内容。

换言之，人们没有感受到地域的变化，不知不觉中，整个场地变成了艺术品陈列室，且整体具有联系性，创造了各种空间，满足了业主的要求。

在这种背景下，建筑师认为他们通过重新思考传统的墙和围栏（限制了区域界线）的概念，同等对待建筑、结构和景观，就能实现一个与城市有新的连接沟通方式的空间。

Atelier-Bisque Doll

Requests and Characteristics of the Site

This is an atelier for a doll artist and a residence of the couple. The request was an atelier that can be used as a gallery at the same time functioning as a doll-making studio. Additionally, it should be a space where their friends, who are often invited, can be gathered around pleasantly. In the residential environment the atelier and the house are wished to be open, while reserving some pri-

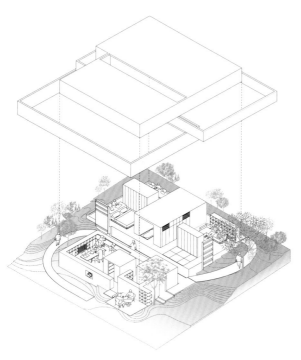

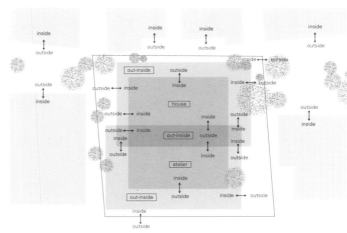

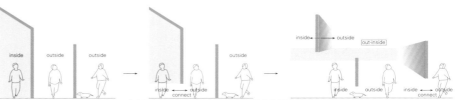

室内与室外空间关系示意图。场地被墙体和栅栏所隔开，
每处空间都改变了分离的墙带与墙体、栅栏和天花板的位置关系。
diagram of the relationship between inside and outside. Fields are separated by walls and fences.
characters of each space alter the states of divided "belts" to the walls, fences and ceilings.

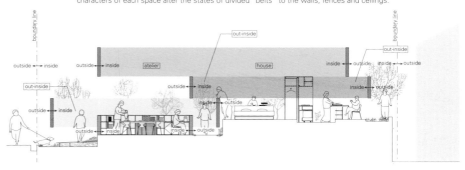

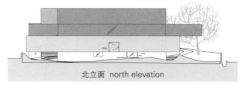
北立面 north elevation

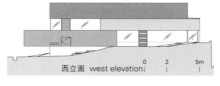
西立面 west elevation

南立面 south elevation

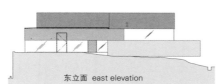
东立面 east elevation

项目名称：Atelier-Bisque Doll
地点：Minoh, Osaka, Japan
建筑师：Keisuke Maeda
结构工程师：Konishi Structural Engineers
机械工程师：K-style
景观建筑师：Toshiya Ogino Environment Design Office
总承包商：Seiyu Kensetsu
用地面积：328.16m² 建筑面积：151.25m² 总楼面面积：151.25m²
结构：steel frame
室外饰面：mortal 室内饰面：mortar, plaster board
竣工时间：2009.11
摄影师：courtesy of the architect – p.37, p.41
©Hiroshi Ueda(courtesy of the architect) – p.30~31, p.32~33, p.34, p.36, p.38, p.39

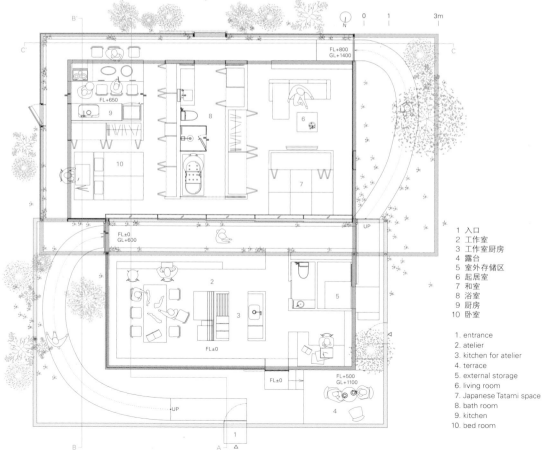

1 入口
2 工作室
3 工作室厨房
4 露台
5 室外存储区
6 起居室
7 和室
8 浴室
9 厨房
10 卧室

1. entrance
2. atelier
3. kitchen for atelier
4. terrace
5. external storage
6. living room
7. Japanese Tatami space
8. bath room
9. kitchen
10. bed room

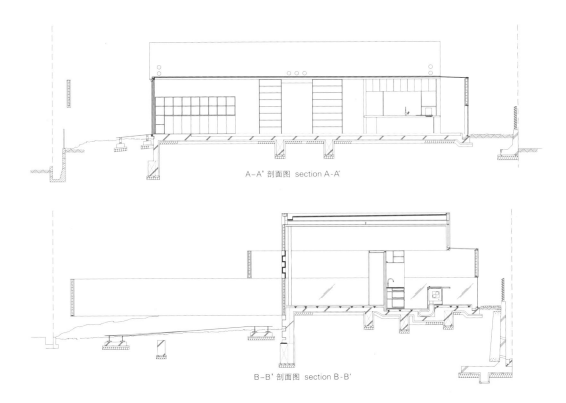

A-A' 剖面图 section A-A'

B-B' 剖面图 section B-B'

vacy from the neighborhood.

When I first visited Minoh City in Osaka Prefecture, I remember the place was plenty of green. It is because the season was spring, but also the city was ordained by vegetation standard, which ordered more than 10% of the site to plant tall, middle and low trees in urban districts with lower than 60% of building coverage ratio.

Form without Territory

In order to secure privacy, we throught if it is possible to do so with more people's involvement rather than building walls or fences on site boundaries and capturing the whole from inside. That is to create the whole from outside by choosing the neighbor's greenery as outdoor space and not producing necessary functions from inside. Not something like walls, hanging walls or fences but waist-high walls that are freed from gravity or floating belts surround the entire site. This is a theory to create architecture that is interior-like extending from outdoor, opposed to substantial in/out-territory.

Floating Belt

Specifically, we expected that the territory, which dose not regulate in/out-terrain, be made by a spatial diversity that is derived from the floating belt, which is accumulated double or triple in rectangle for taking balance, and the belt itself.

Employing 1.2m level of the difference in the site, two functions are allocated to where floating belts overlap. The atelier is put on the lower part facing north when looking front street and the residential piece on the upper south side. An approach was built as a slope as if walking on the slanted topography. By putting a small box that stores necessary functions such as a closet in the atelier and the house, it becomes a space like a street-facing piazza that has no region. The two functions become one room space by triple pile of belts.

A simple operation of overlapping belts that have directionality obscures site boundaries and formulates a relationship of the site and the neighborhood that contains new extent.

That is to say, the entire site becomes a gallery without body-sensed territory and that is connected to create variety of places that the client requires.

In this occasion, we think we could realize a space that has new link to a city by rethinking notion of walls and fences that obstruct boundaries and treating architecture, structure and landscape equivalently. UID Achitects

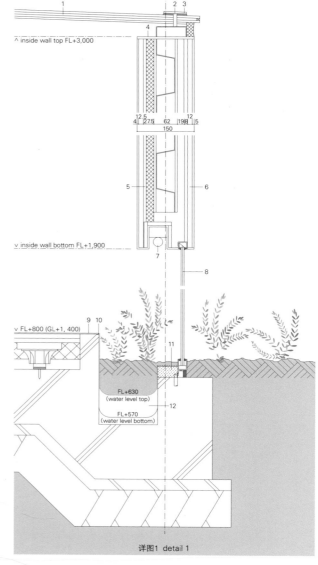

详图1 detail 1

1. skylight: double glazing Low-e HS6+A6+HS6 with protection film
2. stainless rod ø9
3. bead: stainless t4.5 60x60
4. corping: steel flat bar t6 ø150
5. wall: emulsion paint, resin mortar t4, plaster board, insulation spray
6. photocatalyst coating material paint, resin mortar t4, mineral board t8+12, furring strip t19, sandwich panel t62
7. light fixture
8. double glazing
9. floor: polyvinyl chloride sheet t2.5, free access floor, insulation t45, concrete slab t150
10. bead: aluminium angle
11. urethane resin paint mortar
12. mortar drainage slope urethane resin paint

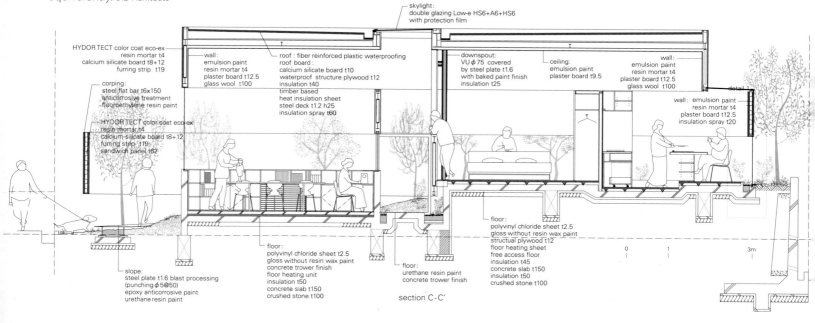

section C-C'

户田屋
Office of Kimihiko Okada

该场地位于广岛市的一处住宅区，建在坡度较缓的山坡上，可以鸟瞰濑户内海优美的海景和整个宫岛的旖旎风光。这个地区发展为具有不同水平高度的平台，业主要求房屋能越过周围较低房子的屋顶眺望到海景，并且要求考虑到安全因素，因为房屋地点位于住宅区的边缘，同时要求预留出扩建空间，使业主能在将来开一家小商店。

为了满足业主的要求，房屋从地面被托起，如同鸟巢一样，让人想到了建筑的首要功能就是消除周围干扰。房屋视野通透，同时给人安全感，远离周围环境。墙板和屋顶都是一体成型的水泥板，形成一个连续的整体。楼梯贯穿整座建筑，使人在屋中自如行走，景随步移，同时又使空间丰富多样。可延伸的金属板使业主未来扩建的心愿成为可能，同时地面层给人的印象也变得柔和。拱肩墙的高度随墙板的厚度发生渐变，两者共同创造了连续但是多样的室内外环境。

建筑底部的洞口使山坡上的微风从此徐徐吹过，螺旋上升的房屋主体形成的垂直空间成为竖向通风风道，为房屋带来清新而自然的空气。细长的形状和悬起的结构使地面层的花园通透、明亮，对附近居民来说就像是一座公园。这样的房屋结构使花园通风和采光都很好，为植物的生长创造了极佳的环境。

项目名称：Toda House
地点：Hiroshima, Japan
建筑师：Kimihiko Okada
结构工程师：Structured Environment
机械工程师：System Design Laboratory
用地面积：189.85m²
建筑面积：90.21m²
总楼层面积：114.26m²
建筑规模：three stories above ground
结构：steel structure
竣工时间：2011
摄影师：©Toshiyuki Yano(courtesy of the architect)

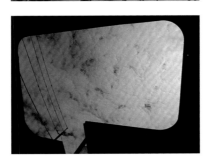
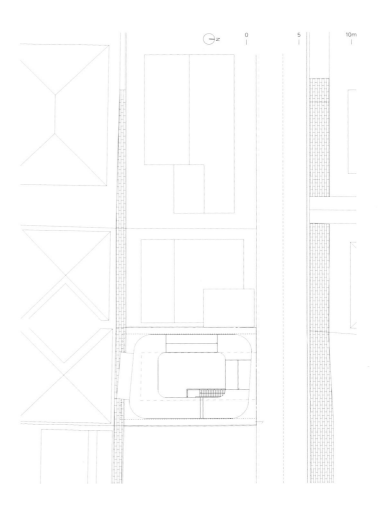

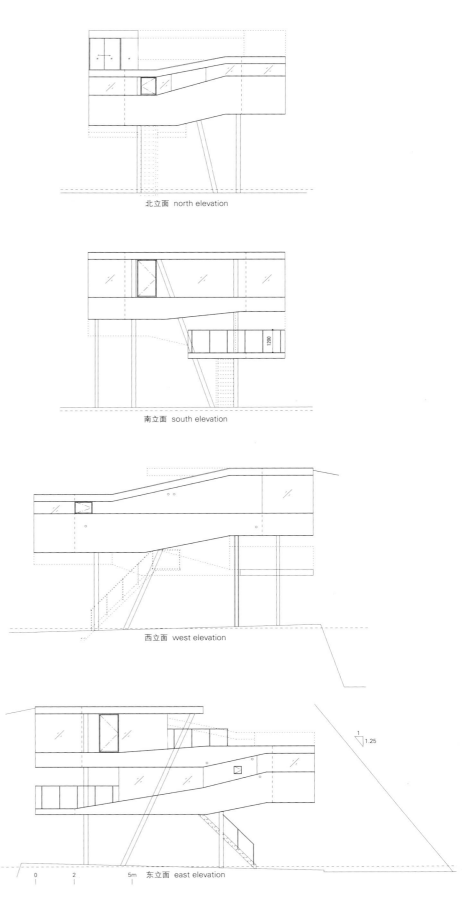

北立面 north elevation

南立面 south elevation

西立面 west elevation

东立面 east elevation

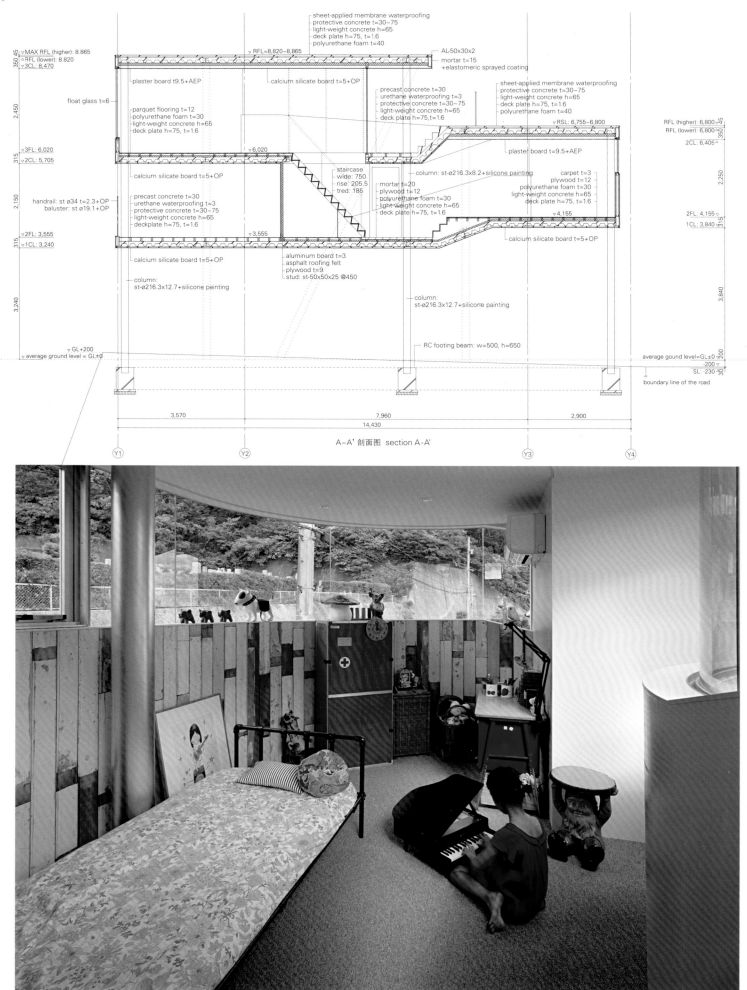

A-A' 剖面图 section A-A'

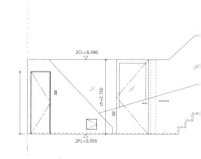

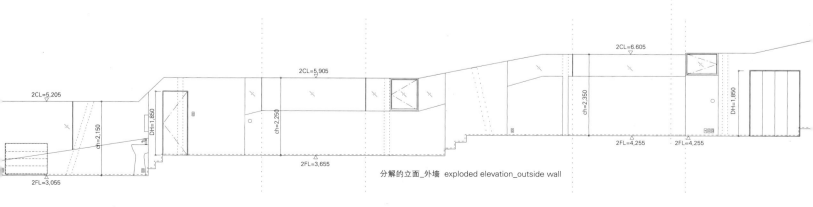

分解的立面_外墙 exploded elevation_outside wall

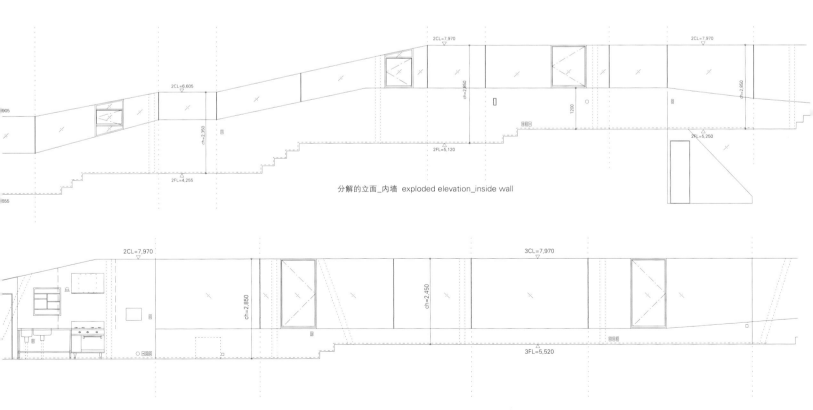

分解的立面_内墙 exploded elevation_inside wall

Toda House

The site is located in a residential area developed on a gentle perch in Hiroshima, overlooking a far view of the Inland Sea and Miyajima. The land of this area is developed into platforms with several levels. The architecture was requested to have a view over the roof of the neighboring house, standing one level lower, and to consider security, for the site is located at the edge of the residential area, and to leave some space for extension when the client opens a small shop in the future.

To respond to the requests, the house was lifted from the ground. Like a bird's nest, it called up architecture's primary function of relief from disturbance. The house was open to the view and yet protected from the fear and environment. Slab and roof consisted of one continuous plate. The variations of circulation and diverse spatial relations were achieved by placing a penetrating staircase. The extended plate made the future extension possible and softened the impression from the ground level. Spandrel wall changed its height accordingly to the thickness of slab. Together with the slab, the spandrel wall created the continuous but various environments.

The opening which catches the breeze that passes through the perch, and the stack ventilation that is operated by the vertical interval of the beginning and the end of the spiral house, give a refreshing natural air conditioning to the house. Because of the elongated shape and the lifted structure, the ground level garden provides an open and bright landscape, like a park, to the neighborhood. This structure also gives good ventilation and light to the garden, creating a rich environment for the plants to grow.

Office of Kimihiko Okada

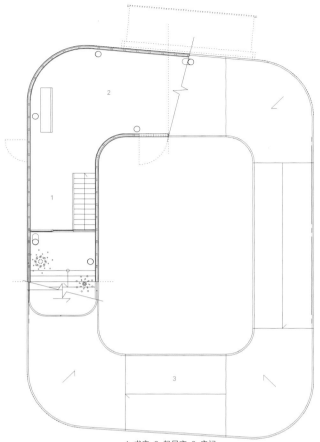

1 书房 2 起居室 3 房间
1. study room 2. living room 3. room
三层 third floor

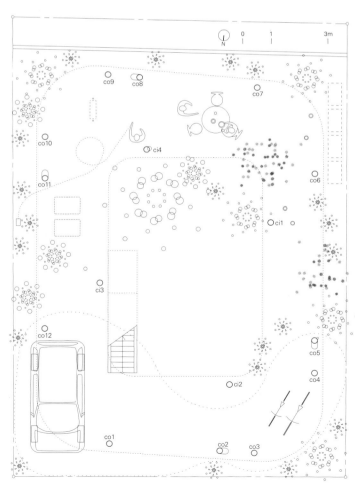

一层 first floor

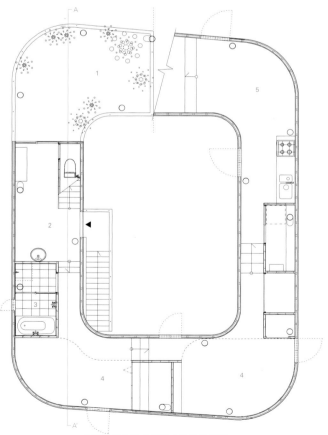

1 露台 2 入口 3 浴室 4 卧室 5 餐厅
1. terrace 2. entrance 3. bathroom 4. bedroom 5. dining room
二层 second floor

Tenjinyama 工作室
Ikimono Architects

春天，花香浸入室内，
夏天，树荫片片洒落，
秋天，枯叶落向重生的温床，
冬天，阳光倾泻，盈满大地。

晴天，天花板倒映着云影，
雨天，天花板高声与水珠相应，
多云，它变成一块抽象的屏风，
有雪，它化为一个纯白柔软的屋顶。

清晨，看着星辰隐没，
白天，感受空气升腾，
傍晚，如血夕阳将室内尽染，
入夜，墙幕上皮影戏栩栩如生。

树木将根深植入泥土，
吸收水分，释放氧气，
人们以此呼吸，以此劳作，
我愿它长大呀，用水辛勤浇灌。

混凝土清减它的僵硬，
花园空间使它温和，
置于变化之中，方知生命转瞬即逝，
时光一旦流逝，昔日寓所也尽成空。

享受常更新的工作环境

这座建筑物的设计初衷就是一个遮风挡雨的地方。

建筑由四面墙和屋顶围合而成，原始的构造表明"一个人似乎怎样都能生活"这一可能性的存在。实际上，这座建筑给人带来了欢乐、奇思妙想和发现。

关于设计方法，其简单性显而易见。打造一个盒形结构住在里面，开扇窗户与小镇保持联系，安装透明天花板以仰望星空，种棵树以避暑纳凉，铺一层土壤供根茎生长，提升屋顶使其伟岸。所以，工作室内外的风景多种多样，超出想象。这种设计带来的舒适感却又不同于身处开放的户外所带来的感觉。

全世界的都处在持续不断的变化之中，地球的自转和公转引起四季、昼夜和天气的变化。如果在城市里，我们处在一个社会，在大自然中，我们被森林、海洋和动植物所包围。自然界的景色是复杂的、美丽的，又是稍纵即逝的。建筑师希望这座建筑会让他一见钟情，又会让他一往情深地把它融于整个大环境中，同时还有小小的新意。这个工作室所处的环境是夏季傍晚多雷雨，冬季天气晴好，但会受到强而干燥的风的影响。

一张2011年冬季的照片记录了Tenjinyama工作室故事的开始，那时，树木还是小树苗。想象一下未来，当孩子已经长大，我们盼望着领略如"出生、抚育、不久消失"这样的场景。

Atelier Tenjinyama

In spring, smell of flower covers the office,
In summer, shade of tree is dropped,
In autumn, the leaf falls in the bed of awaking,
In winter, the sunshine pours into the ground.

On sunny, floating cloud is reflected in the ceiling,
On rainy, the ceiling rings strongly,
On cloudy, it is an abstract screen,
On snowy, it is a white, soft roof.

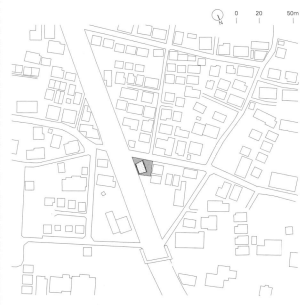

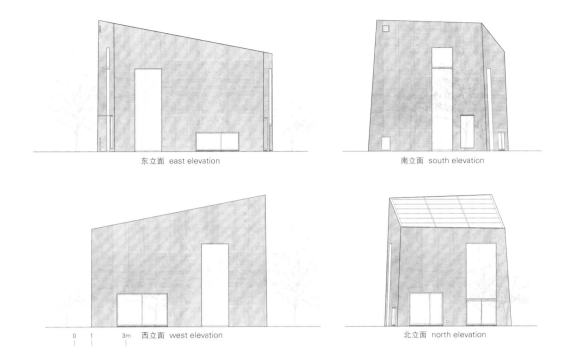

东立面 east elevation   南立面 south elevation

0 1 3m 西立面 west elevation   北立面 north elevation

In the morning, look at a disappearing star,
In the daytime, air rises up,
In the evening, the room is stained with madder red,
In the night, the shadowgraph moves on the wall.

The trees puts the root on the floor of soil,
Resolves water and it makes it to oxygen,
The person breathes in this oxygen and is at work,
To grow the tree, I water it.

Tent concrete is to defend from severity,
Place like garden is to melt to geniality,
To know transience, we are enclosed by the living thing,
If time passes, here becomes a grave.

Enjoy the environment that always changes

This building was born to be able to hold it to prevent rain and wind.

This building is made only on four pieces of walls and roofs. The primitive constitution showed the possibility that" a person seemed to be able to spend somehow". In fact, it rises in joy and presence of mind and discovery here.

As for the design method, simplicity is clear. Make a box to live, establish the window to be connected to the town, grind a ceiling transparently to look up at the sky, plant a tree to make a bower, make a floor soil so that a root grows, raise a ceiling to be brought up greatly. So there are the scenes that are various inside and out, beyond the imagination. This is comfort of the kinds unlike a feeling of outdoor openness. There is comfort to lead to the outside from the place that was followed physically and psychologically, or enjoying wind and rain not to prevent them.

There is our environment that continues changing dynamically around all over the world. There are a season and time and weather coming out of movement of the earth. There is a society if in the city. In nature, we are surrounded by a forest and the sea and creatures. They are complicated and beautiful, and are fleetingly. I want to make the building which I can be impressed by for an instant in an instant entrusting the position to the big environmental flow, and enjoy a slight difference. This atelier is in the environment where an evening of the summer has many thunderstorms, and the winter has fine weather and strong, dry wind blow in.

A photograph of winter in 2011 when trees are still childish, recorded the beginning of the story of Tenjinyama. To image the future when the child is brought up, we wait to enjoy the scene in the future like "Birth and breeding and disappearing before long."

Ikimono Architects

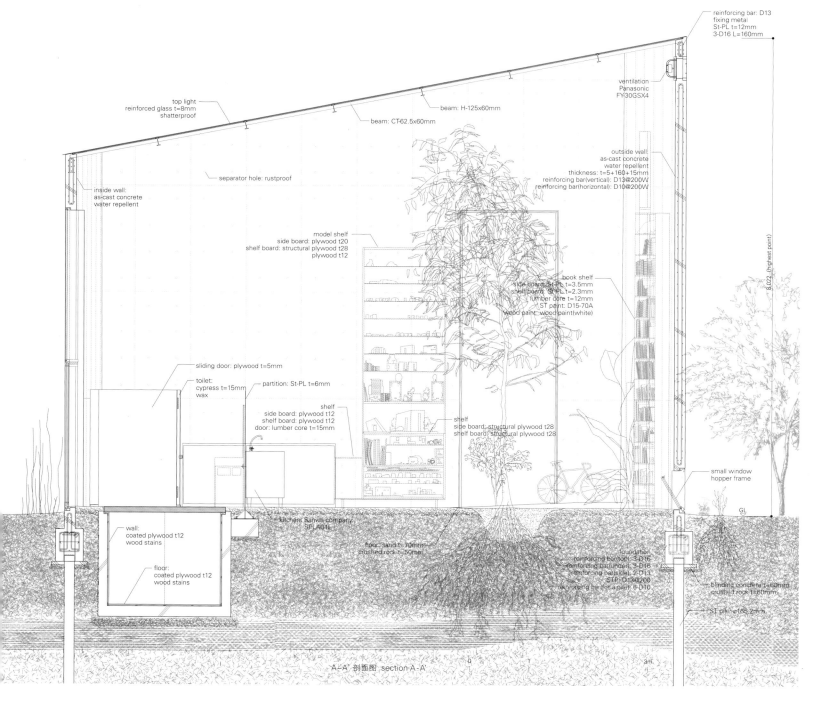

A-A' 剖面图 section A-A'

项目名称：Atelier Tenjinyama
地点：Gunma, Japan
建筑师：Takashi Fujino
结构工程师：Akira Suzuki / ASA
景观建筑师：Atsuo Ota / ACID NATURE 0220
总承包商：Kenchikusha Shiki Inc.
用途：office, residential
用地面积：177.18m²
建筑面积：61.93m²
总楼面面积：61.93m²
结构：reinforced concrete
屋顶：safety glass shatterproof
室外：reinforced concrete
设计时间：2007.7—2010.1
施工时间：2010.1—2011.1

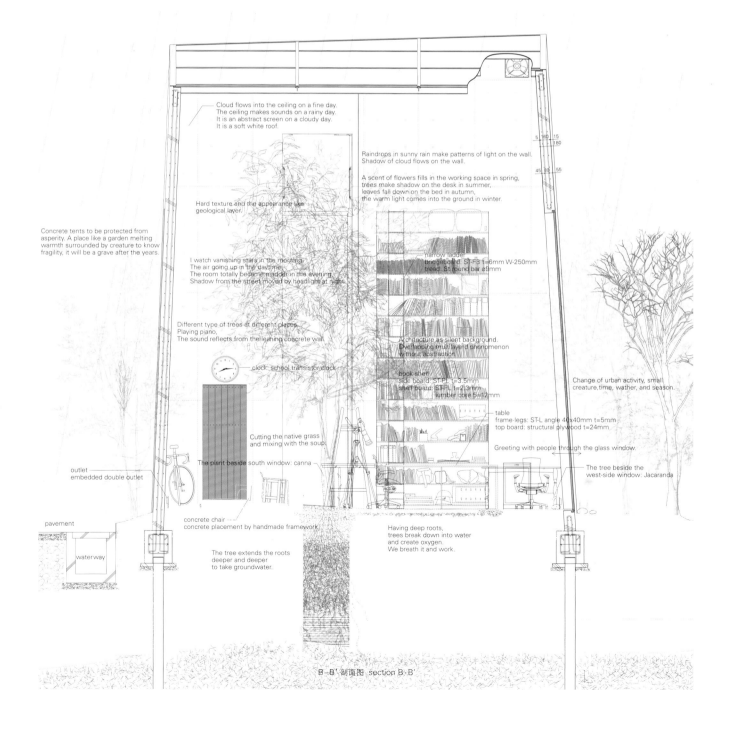

B-B' 剖面图 section B-B'

Ground Folds

大地的皱折

保罗·维瑞利奥于1965年发表了以下言论："根据我最近对三维结构（或悬挑或气动）的研究，本人确信在未来的建筑中将被世人关注的因素不是建筑的外立面或屋顶，而是地面。"

维瑞利奥所发表的声明可以说是对未来三十年的预见，是近期关于建筑学科的最重要的研究之一，即为地面建筑的建设方法下一个定义。事实上，在维瑞利奥提出以上预见的三十年后，伊东丰雄评价说，FOA建筑师事务所在横滨港码头设计的建筑如同是在绘制一种地形。

现代主义运动正处在转折点阶段，该运动关注地面，而并未关注其在建筑物的设计阶段所起到的积极作用。在定义全新的建筑汇语方面的另一个至关重要的阶段是将关于地形的数学性概念和领域的空间这两个因素引入建筑领域。因此，地形学的转变生成了一种关于地表和建筑并置的可行的替代方案。这里所要介绍的建筑都与这几年所进行的试验相关，这些项目都清楚地凸显出大地是如何成为一种地形塑造系统的，该系统摆脱了被动因素的束缚，积极地参与到建筑的建造过程中，从而将它转换成了一处庞大复杂的景观。

In 1965, Paul Virilio remarked, *"I am sure that in the future the dominant architectural element will not be the facade, nor even the roof, as some recent research on three-dimensional structures, suspended or pneumatic, seems to indicate, but the level, the ground."*

Virilio's pronouncement anticipated by approximately thirty years was one of the most important recent pursuits in the architectural discipline: the definition of an operative methodology for the construction of an architecture of the ground. In fact, about thirty years after Virilio, Toyo Ito, commenting on the Terminal of Yokohama by FOA Architects, remarked that their competition plans, rather than representing a level of the building, seemed to portray a topography.

This way was made a definitive break with the legacy of the Modern Movement, which insisted on domestication of the ground rather than the recognition of its active role in the design of the architectural object. Another critical step in defining the new language was based on the transfer to the architectural discipline of the mathematical concept of topology and topological space. The shift towards topological geometry thus generated a viable alternative to the juxtaposition between the ground and the architecture. In continuity with the experience initiated in those years, certain current projects clearly demonstrate how the ground can become a topographical operating system which, freed from its passive condition, fully participates in the configurative process of the building, transferring upon it a greater scale and complexity – that of the landscape.

首尔百济博物馆/G.S Architects & Associates
Monteagudo博物馆/Amann-Cánovas-Maruri
布鲁克林植物园游客中心/Weiss/Manfredi
VanDusen植物园游客中心/Perkins+Will
象征性纪念物/TEN Arquitectos
在大地的皱折里/Marco Atzori

HanSeong BaekJe Museum/G.S Architects & Associates
Monteagudo Museum/Amann-Cánovas-Maruri
Brooklyn Botanic Garden Visitor Center/Weiss/Manfredi
VanDusen Botanic Garden Visitor Center/Perkins+Will
Emblematic Monument/TEN Arquitectos
In the Folds of the Ground/Marco Atzori

在大地的皱折里

19世纪60年代,法国建筑师克劳德·帕伦特介绍了"倾斜机能"的概念,提出如下观点:建筑物与大地之间的传统关系(将竖直的体量置于水平的地表上)受到了质疑,人们可以开创一种全新的理念:通过斜面创造出连续的空间。他主张重新创建一种流畅的关系,把建筑物的所有空间视为一个整体,在这个空间里,用户及其交通流线以及机械循环系统都是以一种连续的方式布局的。因此,帕伦特彻底改变了人们对大地和建筑两者之间的双重性的认知,并开创了一个崭新的研究领域,在接下来的几年中该领域逐渐发展成熟。事实上,在后来的几年中,克劳德·帕伦特的这种认知已经融入到了一种更为宽泛且复杂的理论框架中,该框架旨在创造出一种建造大地和景观的方法,并界定一种混合的调查领域,其中包括建筑与景观,甚至是建筑与几何。

这种学科环境已于20世纪90年代成形,该学科将景观和地域视为不断建造和加工的产物,并认为建筑的宗旨就是在不同规模的元素之间建立起一种联系,从而打造出一种更加流畅、复杂的逻辑,这一点还要归功于地理空间。这样一来,建筑师们就优化了将建筑融为景观的这一理念。土地不再是支撑建筑的被动因素,而是比空间构建更具主动性的存在,成为重新构建水平表面的新方法论的主要因素。

这项研究取代了传统的建筑理念,大地成了研究空间的场所,而不再是项目可以忽视的元素。在这种情况下,比起仅用高度来测量的土块,把各种表面相互重叠而造出一个新的"地形"变得尤为重要。Yorgos Simeoforidis对此评价说,"在这种新的构造中我们能够发现易逝的线条、模糊的形状、内外的连续。这种由里向外所构成的建筑可以与自然相结合,从而打造与灵活'地形'相匹配的空间。"

从数学的角度来说,如果地形学是对形状和形式的研究,或者换句话说,如果地形学定义了一系列规则,而这些规则又定义了空间元素之间的关系、联系和连续性,那么从地域性角度来说,地形学在形态学方面可以被归为景观学。

In the Folds of the Ground

The French architect Claude Parent introduced in the 1960s the concept of "oblique function", through which he argued as follows: faced with a crisis in the traditional relationship between the architectural object and the ground, as witnessed by the prevalence of the vertical dimension (volume) over the horizontal (ground), one could envision a new concept – that of continuing the space, with the oblique dimension generating continuity with the built space. Parent's thought appeared to unequivocally necessitate the reformulation of a fluid relationship which reorganizes the perception and use of space on a global scale, in which human occupation and pedestrian and mechanical circulation are possible in a continuous manner. Parent thus profoundly changed the perception of the duality between the ground and architecture, and introduced a new research field which would reach full maturity only in the following years. In fact, in subsequent years Claude Parent's intuitions would be incorporated into a broader and more complex theoretical framework intended to build a methodology for manipulating the ground and landscape and identifying a hybrid field of investigation consisting of architecture and landscape, or even architecture and geography.

This disciplinary environment, which would take full shape in the 1990s, considers the landscape, the territory, as the result of continuous processes of manipulation and processing, and considers the vocation of architecture to be the building of relationships between elements at different scales to introduce a more flexible and hybrid logic that is increasingly attributable to the geographical dimension. In this way architects have refined the idea of an architecture conceived as landscape based on the reconfiguration of horizontal surfaces into the main elements of an operative methodology that will impart to the ground, until recently considered a passive element providing support for a building, an active role in the construction of spaces.

As an alternative to the traditional conception of architecture, the research concerning the ground becomes an investigation of the vacuum rather than that of an object to which the project does not conform. What emerges is the configuration of a mass built not only in height, but through the creation of a new "topos" created by an architecture made up of overlapping surfaces. Yorgos Simeoforidis refers to this approach as *"new mechanisms prone to be evanescent silhouettes, vague shapes, a fluid continuity between interior and exterior space. An architecture made from the inside, in communion with nature, capable of generating spaces elastic and flexible, definitely topological."*

If topology, in mathematics, is the study of the properties of shapes and forms, or in other words, if topology is what defines the set of rules that explicitly define the relationship, the connection and continuity between spatial elements, from a territorial point of view – the concept of topology can be assimilated into the study of landscape from a morphological standpoint.

By this logic, the construction of a project involves interpreting

横滨国际码头,日本,由FOA建筑师事务所设计,2002年
Yokohama International Port Terminal, Japan, by FOA, 2002

按照此逻辑,项目的建造包括诠释在地球上存在的自然和人工的空间符号,从而将它们转换成有关土地的新语言。另外,这样的结果使我们可以设计出更细腻、更流畅的空间。德勒兹提出的"皱折"是定义三维空间的元素,也是之前界定体量不可或缺的因素。

最终,表面和体量区分的边界变得模糊,我们在大地的皱折间所产生的空间里规划。建筑颠覆了其传统概念,以强调能够从建筑内部观赏到的周围景观,而不是将建筑自身作为周围环境中的主要结构。

在对大地潜能进行过潜心研究的建筑师当中,亚历杭德罗·柴拉·波罗与Farshid Moussavi(之后这二人成立了FOA建筑师事务所)在一篇名为《大地的再整形》的文章中,清晰地阐述了他们所研究的领域:"我们所设计的项目是在探索如何利用更多样的方法和技术进行再创作,而不是关注土地的稀缺,这将是一门崭新的土壤学。人类对于大地表面的改造在不断地进行中,也正因如此,很容易将定数转变为积极的变量,变得复杂,也会产生突变的情况。这些改造从在地面建立起现代化的高楼大厦,到使被人类野蛮改造过的大地恢复生机,不一而足。"

这种观点和帕伦特的一些概念也是不谋而合,例如为了创建出空间的持续性而利用斜面这种方法。然而,在这基础上,FOA建筑师事务所针对土地这一焦点要素进行研究并提出了一些启示:"从物理层次或文化角度来看,大地是非自然产生的人工建筑物,虽然它既不是主题也不是背景,但可以启动某种物质的基本系统。风景在这样广范围的意义下进行理解的时候,它的确是一个奇妙的存在,即土地是创建新地形的系统,并非是构建环境所形成的众多要素之一,它本身只是一个广阔的平台,而不仅是一块地皮。"

如果要重新对概念结构下定义,必须从以下几种角度来思考。我们需要做的就是必须立足于如何重新针对土地的地形下定义,而不是一味地研究如何构建建筑物。这些方式能够引导设计师创建出一种全新的结构形式,在这种形式中,静态符号将被图像、图片以及动态模拟装置所取代。平面和人工地形或者标注某些区域的地图相似,截面看上去与断层一样。

the signs, natural and artificial, which overlap the area, converting these into referents of a compositional language associated with the project of the ground. The spaces are then designed in smooth and supple continuity without breaks. The fold (also in the philosophical declination Deleuze confers upon it) is the element that determines the definition of three-dimensionality and therefore of the emergence previously identified in the volume. Consequently assuming a paradigmatic value are the perceptive ambiguity between the surface and the volume that it determines, as well as the programmatic space generated as an interstitial element between the folds of the ground. It's an architecture that subverts its conception so as to emphasize the surrounding landscape as observed from within, rather than insisting upon itself as the main object of observation from outside.

Among the architects who have most deeply investigated the potential of the ground, Alejandro Zaera Polo and Farshid Moussavi (then better known as the Foreign Office Architects), in a text entitled "The Reformulation of Suelo," clearly explained the areas which moved their research: *"Our project on the surface does not reflect on the absence of the soil, but on the redefinition and the creation of a number of techniques: a new discipline of soil. The manipulation of the surface of the ground has been a constant, which has transformed an element that usually led to a fixed code into an active field, complex and mutant – from the taming of the ground that characterizes modern architecture, to the recovery of a potentially savage differential intensity."*

In FOA's reflection one may discern aspects of Parent's concept, such as the use of the oblique dimension as a tool to create spatial continuity, but FOA also define additional points as foci of research. They envision these new grounds as *"artificial constructions and not natural spaces, both physically and culturally. They are neither figures nor backgrounds, but operating systems, and the landscape is interesting only if it is viewed in the broader significance: as a kind operating system topography, and not as a category of the built environment, a platform and not a site."*

The redefinition of conceptual structures, the introduction of topology as a geometric system of reference, the idea that the project is based on the redefinition of the topography of a place rather than the introduction of an object within a site – all lead designers to a new type of representative and then constructive structure wherein the static notations give way to such tools as diagrams, graphics, and dynamic simulations. The plans move closer to cartographic representations of artificial topographies or territorial maps, while sections appear as stratigraphies.

All these inquiries pave a road that, while prone to diversions in the following years, has influenced later generations in a profound way, to the point that the projects that follow can be considered contemporary interpretations of the operative methodology con-

虽说这个新理论在日后研究中没有取得长足的发展，但直到今日众多建筑师仍深受其影响。下面要介绍的实例就是对当时把土地这一因素引入的方法理论，利用现代观点进行重新讨论。

首尔百济博物馆_G.S联合建筑师事务所

G.S联合建筑师事务所将他们的这个项目评价为"一座公园式博物馆"，这会让人联想到FOA建筑师事务所曾经将他们设计的横滨码头称为"一个公园式码头"。此种描述带来了某种关于土地学科的普遍性假设，即感知方面的模糊性以及空间和功能方面的连续性。这时，地形在功能和空间这两个层面里，并不能区分自然因素和人工因素，而是令两者合二为一的工具。事实上，如果明确地区分内部和外部，或者建筑和承载建筑的大地，那么就没有必要关注任何建筑要素的连续性了，建筑的屋顶也同样可以偏离所谓的大地延伸的地理价值。相反，这里提到的首尔百济博物馆建筑却明显运用了构建地形的设计方法。土地与覆盖层间的连续性造就了一种有形的建筑体量，好像这座建筑就是大地的一种变形，又好像一种显示出地质剖面的结构。建筑体量被设计成地壳内层的样子：扭曲的形状、倾斜的平面、不平衡的结构。这里的建筑与其说是构建独立个体，不如说是这块土地的形态变化。相同的室内空间与室外相连贯，好似延伸进了构建的土块中，从而建成了这个博物馆项目。

布鲁克林植物园游客中心_韦斯/曼弗雷迪

由韦斯/曼弗雷迪建筑事务所设计的"布鲁克林植物园游客中心"最显著的部分是其宽阔的屋顶，凸显了植物园（自然）与城市（人工）之间的关系。建筑师对地表的处理创造出了一种当地环境，在这里，我们发现了城市与自然的规模，这些都反映在了选材和地表开发方面。通过设置能够标记出下方路径的百叶窗，面朝城市的屋顶花园可以与建筑外表最坚固的部分相关联。而其下方的空间才是该项目真正的所在之处，该空间是一个几乎完全透明的体量，一个采用了现代技术的温室，为叠加的地形起到了支撑作用。这座具有传统理念的建筑不得影响到周边景观的结构。该体量是一个小型的基础设施，人们可以在这里探索并理解外界的空间。韦斯/曼弗雷迪始终将建筑与环

cerning the ground which emerged in those years.

HanSeong BaekJe Museum _G.S Architects & Associates

"A museum like a park", so G.S Architects & Associates say of their project, a sentence that recalls the FOA's description of their Yokohama terminal: "A terminal like a park." Such descriptions bring with them certain general theoretical assumptions concerning the discipline of the ground, namely perceptive ambiguity and spatial and functional continuity. The topography is the tool for artificially modeling a built space in which there is no distinction between the natural components and artificial light, both in the compositional strategy and in that programmatic. Wherever there was a categorical distinction separating the inside from the outside, or the architectural object from the ground on which it rests, the tectonic elements cannot merge with continuity, nor can the roof of the building assume the value of a topographical element in continuity with the ground. By contrast, where the architecture, as in the case of the HanSeong BaekJe Museum, assumes the meaning of a topographical operational tool, the continuity between the ground level and the level of coverage results in a building volume that materializes as if it were itself a movement of the ground, as if a telluric action had brought to light a geological section. The volumes of the building seem inner layers of the Earth's crust come to light: the shapes are distorted, the planes are oblique, the structures seem in precarious balance. The building seems the tale of the morphological modification of a place rather than the composition of an architectural object. The same interior space, in continuity with the outside, seems dug into the emerged masses, hosting the museum program.

Brooklyn Botanic Garden Visitor Center _Weiss/Manfredi

The Brooklyn Botanic Garden Visitor Center by Weiss/Manfredi Architecture finds its element of greatest intensity in the creation of a cover that takes care to emphasize the relationship between the garden (nature) and the city (artifice). The manipulation of the topological surface coverage creates a local condition in which we find the scale of the urban and the natural, didactically represented in the choice of materials and in the topological development of the two parts. The construction of the roof garden is related, through a brise-soleil that marks the path below, to the hardest part of the cover, oriented towards the city. The space below, intended to receive the program, is an almost completely transparent volume, a contemporary technological greenhouse which seems to be a support for the overlying topography. The building (the architecture in its traditional conception) is here hierarchically subordinated to the structuring of the landscape. The volume intended to receive the program is a minimal infrastructure that allows one to explore and understand the outer space. Weiss/Manfredi, whose inquiries place the relationship between architecture and environment at the center of their thought, de-

海滨青年之家，哥本哈根，由PLOT设计，2004年
Maritime Youth House, Copenhagen by PLOT, 2004

境之间的关系铭记在心，并将布鲁克林植物园游客中心定义为"花园和城市之间、文化和耕种之间的一条界线"，同时在理解景观时将建筑当作一种中介、一种过渡的工具。据韦斯/曼弗雷迪所说，对建筑与场地之间的关系更进一步的变化"是将该中心设计为一个花园路径系统的三维延续结构，同时框定出一系列花园中的风景"。

象征性纪念物_TEN建筑事务所

对大地进行塑造可以成为一种界定项目的全新工具，这一理念在TEN建筑事务所设计的"象征性纪念物"中发挥得淋漓尽致。与先前的几个项目（由OMA设计的阿佳迪尔会议中心和横滨论坛、由BIG+JDS设计的海滨青年之家以及由FOA设计的横滨港口）相比，象征性纪念物这一项目除了将这座纪念性建筑设计成一个向城市开放的公共空间之外，还建造了一种连续不断的皱折式人工地形，上面承载着功能性开发项目。皱折和延续性可以作为一种将表面转换为体量的方法，该方法在之前提到的项目中都得到了应用。这样的构造结果就是形成一种新型的城市空间结构。同G.S和韦斯/曼弗雷迪设计的项目相比，此作品效果更明显，探索了大地（支撑建筑的场地）与体量（建筑的规模）之间的感知模糊性。通过地形连续性创造出来的空间形成了一种介于体量和表面之间更为深层的联系，在这种关系中表面不再仅仅是一种围护空间的外壳，而是空间形成的一个决定性因素。最终形成的建筑将是一个未界定的模糊结构。因此，大地摆脱了被动因素的束缚，积极地确定了一种全新的规划过程。

VanDusen植物园游客中心_帕金斯+威尔

这座由帕金斯+威尔设计的VanDusen植物园游客中心采用了生态技术。在某种程度上，该项目与韦斯/曼弗雷迪设计的项目有异曲同工之妙，都追求地形的概念化，并更加注重可持续性。事实上，其塑料外壳反映了建筑师想要创造一种复杂流畅的屋顶花园的想法，以弥补建筑给周边景观带来的影响，从而在必需的人工结构和现存环境之间重塑一种平衡。在探寻共生的可能性的过程中，建筑师受到了当地植物的启发，设计出了该建筑的外形，同时建筑构件以及技术系统的选择都是为了尽量减少对周边环境的影响。为了达到这一目的，建筑

fine the BBCVC as *"an interface between garden and city, culture and cultivation"* with an emphasis on the building's role as a mediator, a dimension of transit in the understanding of the landscape. A further variation of the relationship between architecture and place according to Weiss/Manfredi is that *"the center is experienced as a three-dimensional continuation of the garden path system, framing a series of views into and through the garden."*

*Emblematic Monument*_TEN Arquitectos

The idea that the modeling of the ground can be a new operational tool to define a program is well represented in the *Emblematic Monument* by TEN Arquitectos. In a historical comparison with previous projects that range from the Convention Center in Agadir to the Yokohama Forum (both designed by OMA) and from the Maritime Youth House by BIG+JDS to FOA's Yokohama Terminal, the Emblematic Monument, in addition to providing an alternative view of the concept of the monument as a public space open to the city, has built an artificial topography that hosts the development of the functional program in the continuity of its folds. The value of the fold and the continuity as a method of transforming a surface into a volume is in line with the above-mentioned projects, applied with ability and pragmatism. The result is a new kind of space within the urban fabric, which, with even greater intensity than the work of G.S and Weiss/Manfredi, explores the value of perceptive ambiguity between the ground, intended as a place dedicated to the support of the architecture, and the volume as a representation of the "noble" dimension of building. The creation of the space through topographic continuity develops a deep connection, almost an intimate relationship, between volume and surface in which the surface is not merely the shell that encloses the space, but also its determinant. The resulting figure is an undefined and ambiguous element. It is the ground which, freed from its passive condition, has actively determined a new type of figurative process.

*VanDusen Botanic Garden Visitor Center*_Perkins+Will

The Van Dusen Botanical Garden Visitor Center by Perkins + Will is an eco-tech variant of the reasoning leading to this moment. In some ways similar to the work of Weiss/Manfredi, it pursues topological conceptualization as a minor element and pays greater attention to the components of sustainability. The figuration of the plastic shell, in fact, responds to the desire to create a complex and fluid roof garden to compensate for the presence of the building in the landscape, thus restoring a balance between the necessary artificial component and the existing environment. In the search for possible co-existence, the form of the building is inspired by the vegetal elements typical of such places, and the construction elements and technological systems are chosen to minimize impact. The compositive choice of the covers, to that end, enhances the ability to control, by means of passive systems, the internal microclimate, and the roof garden supports technological systems to capture light and for the production of energy, including via

师通过利用被动系统以及室内微气候等方法，使覆层可以增强控制能力，屋顶花园支持技术系统，以收集光照和生产能源，其中包括利用光伏系统。如果之前介绍的项目都是以理论为主导，那么帕金斯+威尔设计的VanDusen植物园游客中心则可称得上是可持续性建筑的实例，因此与之前的项目有所不同。在该项目中，地质和地形起到了重要作用。然而，地形也限制了建筑的几何形式，与其他项目相比，它好像并不是旨在探究自然与建筑之间的关系。

Monteagudo博物馆_阿曼−卡诺瓦斯−马鲁瑞

阿曼−卡诺瓦斯−马鲁瑞设计的Monteagudo博物馆并未太强调地形这一概念。如果说先前的项目主要强调的是大地与建筑的连续性，那么这个项目则大不相同，它将场地的地形与建筑之间的关系当成了一种地形系统。该建筑的形状通过连续性而得到了凸显，这种连续性反映在了立面和屋顶的选材上。该建筑的形状与周围的地势相似，逐渐延伸至博物馆。该体量通过扭曲的几何形状而得到了界定，采用了一种分段式组装技术，并最终形成了一个完整的结构。如果地域和景观揭示了各个组成部分的内在联系，那么建筑本身同样可以成为一个讲述故事的景观。建筑师称，"这栋建筑无疑是一只颜色与形状相混合的寄生物，植物覆盖了整座建筑，凸显出其能够融入周围环境的能力"。在这个由阿曼−卡诺瓦斯−马鲁瑞设计的项目中，将建筑设计得与周边环境相似并不意味着拟态或伪装。与之相反的是，相似的元素间有一种紧绷的关系，而建筑与地形成了密友。从这个意义上看，建筑师为建筑物选择柯尔顿钢板作为覆层显得合情合理，因为与周围群山的岩石峭壁相比，建筑给人们的感觉是和谐统一的。

在当代建筑学上，语言、理论的增长与迅速分化正在削弱建筑师的地位。为了掩饰这种趋势，对于将大地作为一种操作方法所进行的反思（这种反思造就了本文所介绍的五个项目）可以被看作是现代建筑的重要文化基础，这种方法可以创造出新空间、新景观、新结构、新地形，人文与自然之间的关系已经使文化基础发生了改变。

photovoltaic systems. If the works outlined above are characterized by a strong theoretical component transferred to the project, the work of Perkins+Will for the VanDusen Botanic Garden Visitor Center distinguishes itself instead as a set of good practices for sustainable construction within which the topographical and topological components in the definition find an important space. This component, however, which also exercises control over the geometry of the building, does not seem, by contrast with the other cases, to aim to explore the conceptual dimensions of a possible hybridization between nature and architecture.

Monteagudo Museum_Amann-Cánovas-Maruri

The Monteagudo Museum of Amann-Cánovas-Maruri introduces, in its turn, a possible declination of the concept of topography. If in the previous cases the buildings developed the possibilities introduced by the continuity between ground and cover, in this example the approach is dissimilar, moving on the relationship between the topography of the place and the architecture as a topographic system in its own right. The shape of the building, highlighted by the continuity in the choice of the covering material of the facade and roof, is an assimilation of the surrounding topography which extends to the conception of the museum's masses. The volumes are defined via distorted geometries, using a technique that evokes an assembly of fragments, each with an identity which contributes to the overall construction of a story. If the territory, the landscape, can be understood as a "discovery" of the intrinsic connections between its components, it likewise involves the building, which can render itself a storytelling of the landscape. The architects say that *"this building is, without doubt, a parasite, highlighting the ability of the building to assimilate the conditions of the context as a parasite, [with] blends of colors and shapes and covered with calligraphic plant skin lining the entire building"*. In this operation by Amann-Cánovas-Maruri, the assimilation of the conditions of the place thus does not imply mimesis or camouflage. Rather, the relationship is defined as a tension between otherwise similar elements, as a result of which the place has an alter ego to deal with. In this sense, the choice of the coating material, the corten-steel that characterizes almost the entire building, is logical, because it defines a unitary perception that compares itself to the rocky cliffs of the surrounding hills.

In contemporary architecture the proliferation and accelerated stratification of languages and theories is leading to an annulment of the critical position of the architect. Belying that diffusion, the reflection that has structured these five projects, of the discipline of the ground as an operating method, can generally be regarded as an essential cultural matrix of contemporary architecture, able to produce new spaces, new landscapes, new structural conformations, new "geography" of places, in evidence of a relationship between humanity and nature that has altered its cultural foundations. Marco Atzori

大地的皱折 Ground Folds

首尔百济博物馆
G.S Architects&Associates

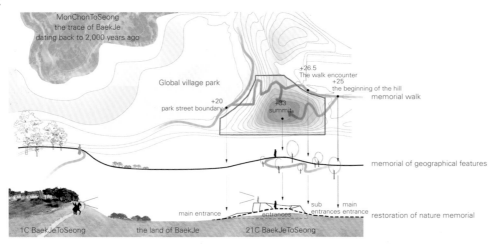

公园式博物馆

首尔百济博物馆位于首尔（原来的汉城）的一片曾经长满松树的低矮小山坡上，与梦村土城相望。站在奥林匹克公园内放眼望去，该山坡呈现出圆滑的曲线，是一道亮丽的风景。此外，山坡顶部高于其周围约13m，站在上面可以向北眺望梦村土城。如果为了建造博物馆而让这片古老的小山坡消失的话，那么对于每个人来说将是一件非常遗憾的事。

如果博物馆的本身就是山坡会怎么样？镶嵌在自然地形中，融进山坡结构，被公园和散步的小路所围绕，难道不是理想的设计思路吗？为了实现这一理想画面，建筑师想到了古代百济时期的土城。

百济土城——博物馆的设计动力

虽然汉城作为首都的百济时代拥有500年的历史，但与泗沘作为首都的百济时代相比，留给我们的印象却是微乎其微的，只有石村洞、梦村土城和风纳土城的古墓群被岁月风化了1500年之久，以真实的面目展现在我们面前。但我们从中还是会发现作为汉城百济时代的历史遗址中具有代表性的梦村土城和风纳土城同高句丽在汉江对面的峨嵯山建造的土城，或者新罗王朝建造在山上的城堡在建筑方式上存在着明显的不同。新罗或高句丽时期的城堡都是采用石头建成，而百济的城堡却是用泥土一层层砌成的。于近处观察，土城的横截面会显露出压实的泥层（垂直向）以及堆积的泥土层（水平向），这构成了用以体现汉城百济时代的最佳建筑主题。

百济时代的地下汉城

这座博物馆的形态可以说是小山坡，也可以说是泥土砌成的城堡，建筑师赋予了它外观上的模糊性。展览室呈现泥土城堡的形态，环抱着场地的土坡，同时保持着土坡的高度，利用下边的地下空间来容纳博物馆大厅。展览空间位于地下，不仅仅是因为建筑师希望该结构成为土地的自然延伸，还想让其成为"地下的百济"（百济时期的汉城通过考古发掘所展现出来）的具体体现。

百济遗址的美景

泥土城堡和山坡必然要方便让人们攀爬，所以泥土城堡的展厅屋顶设计成斜坡，使人们更方便登高散步。博物馆屋顶的漫步小径与奥林匹克公园的散步甬路相连接，不仅构成了公园的一部分，而且还成为城堡的流线，为人们提供了在古代百济大地上徜徉的机会。小径的尽头就是和山坡（一直存在的）等高的展望台，站在这里眺望，可将对面百济人民当年用鲜血和汗水筑成的梦村土城尽收眼底。

博物馆设施

博物馆在一楼和地下分别设有两个出入口。为了可以进入泥土城堡展厅的主大堂，即博物馆的主要部分，人们需要沿着小路顺势而下。通过入口人们进入到教育中心。此外，二层还设有工艺品店、特殊展厅以及一个常设展厅（展现了汉城百济以前时代的历史）；三层全部用作展示整个汉城百济历史的常设展厅；四层则设有小型自助餐厅。此外，地下一层和地下二层设有研讨室、图书馆以及能够容纳130个座位的表演大厅、储藏室和停车场等。

风纳土城

博物馆的主要特色就是复制了风纳土城的墙体，其底部长43m，顶部长13m，高10.8m。为了让这面巨大的墙壁一目了然，从地下层贯穿至地上二层的主大堂全部开放。

展厅沿着防御土墙设计为圆形，随着时间的流逝，墙体的高度也随之变高，末端最高的部分就是高潮内容，即展示汉江区域三国争霸的空间。人们重温这片区域的历史记忆，再现古代百济人们骑马奔驰的故土。

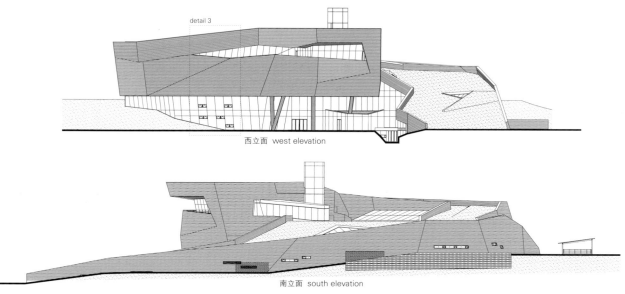

西立面 west elevation

南立面 south elevation

HanSeong BaekJe Museum

Museum like a Park

The land on which HanSeong, present-day Seoul, BaekJe Museum is located used to be a small and low-rise hill with pine trees. The hill that faces MongChon ToSeong(Mud Castle) was seen to create a beautiful landscape shaped by round and curvaceous land viewed from Olympic Park. In addition, the peak was about 13m higher than its surroundings and was good for northward observation of MongChon ToSeong standing on the top of it. If the long-standing hill had disappeared to build a museum, it would have been sad for everyone.

What if a museum could be a hill itself? Would it not be ideal to have the building buried in the natural terrain thus becoming the hill, surrounded by a park and a footpath for a leisurely stroll? To realize this ideal image, we thought about the mud castle of BaekJe.

Mud Castles of BaekJe – A Design Motive for the Museum

Though HanSeong BaekJe era lasted for 500 years, there is not much known to us, compared with the SaBi era. Only the cluster of the old tombs located in SeokChon-dong, MongChon ToSeong and PungNap ToSeong have weathered for 1,500 years, presenting us with their real contents. The MongChon ToSeong and PungNap ToSeong, the representative historical sites of the BaekJe Dynasty in HanSeong were built in an architectural method conspicuously different from what fashioned the castle KoKuRyeo Dynasty built up on Acha mountain across the Han River or those of the mountain castles of the ShiLa Dynasty. While ShiLa and KoKuRyeo made their castles with stones, BaekJe stacked layer upon layer of mud to build theirs. A close look at the cross-section of the mud castle reveals a vertical stack of compacted mud layers, and the horizontal layers of soil piled up. Thus constituting the most excellent architectural motif that expressed the BaekJe Dynasty in HanSeong.

HanSeong BaekJe discovered underground

We decided to give the museum a formal ambiguity by letting it serve as a hill and a mud castle. The exhibition room is to take the form of a mud castle and encircle the mounding on the site, and while the height of the mounding is kept intact, the underground space is emptied to accommodate the museum lobby. The exhibition space was placed underground not just because we wanted

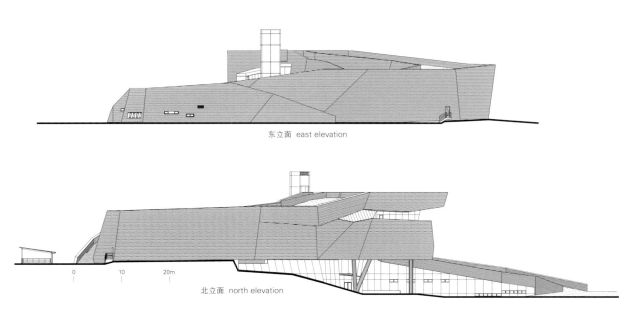

东立面 east elevation

北立面 north elevation

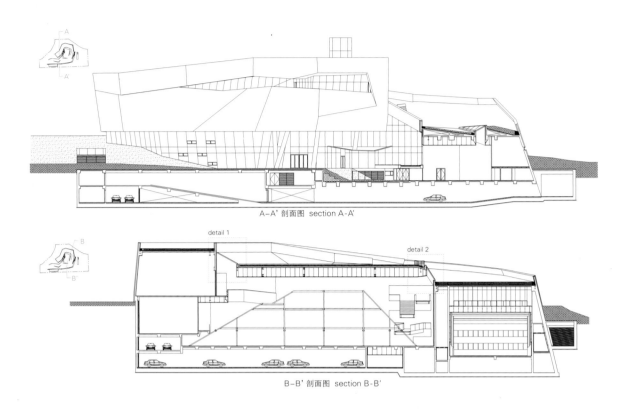

A-A' 剖面图 section A-A'

B-B' 剖面图 section B-B'

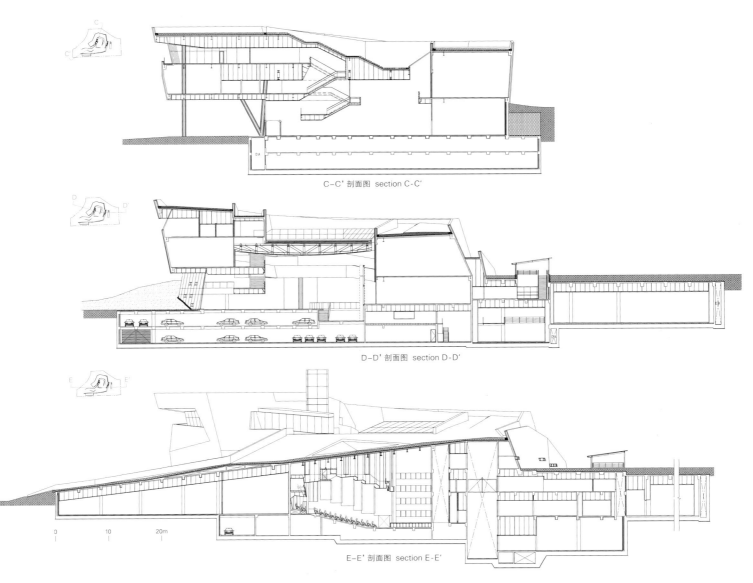

C-C' 剖面图 section C-C'

D-D' 剖面图 section D-D'

E-E' 剖面图 section E-E'

the structure to serve as natural extension of the land formation, like a mounding, but also it was intended to be a metaphor for "BaekJe discovered underground" in that the reality of the BaekJe Dynasty in HanSeong is being unearthed through the archaeological dig.

A Nice View of the Historical Site of BaekJe

A mud castle and mounding must be climbed by people. So the rooftop of the exhibition room has been designed as a slope so that folks can take a walk by climbing it. The walkway over the museum rooftop leads to the trail in the Olympic Park not just to make it be a part but also to form a route around the castle, providing an opportunity to tour the old domain of BaekJe. The end of the walkway is level with the mounding that has been there all along, naturally creating an observation deck, where people stand vis-à-vis MongChon ToSeong, which was created by the sweat and blood of old BaekJe people.

Museum Facilities

Upon approaching the museum you will find two entrances, one located on the ground floor and another in the basement. In order to gain access to the main lobby for the castle exhibition hall, the main part of the museum, one needs to use the walkway that leads downwards. Through this entrance access to the education center is also granted. Additionally, there is an art shop, a special exhibition hall, and a permanent exhibition hall which showcases the era of pre-HanSeong BaekJe on the first floor. The entire second floor is being used as a permanent exhibition hall for the HanSeong BaekJe era. Lastly, on the third floor you will see a small cafeteria. Also, there is a seminar room, a library, and a performance hall which has 130 seats, preservations and parking on the first and second basement.

PungNap ToSeong

The main figure in the museum is the wall which replicates the PungNap ToSeong. Its bottom length is 43m, top length is 13m, and the height is 10.8m. To offer a view of the huge wall, the main lobby is open widely from the basement to the second floor.

The exhibition hall is shaped like a circle around the rampart. As time goes by the wall becomes higher and higher. At the top of the wall, you will see the climax, the space showing the fight for supremacy in three nations around the Han River. We can relive the historical memory of the area as the place where old people were riding horses. G.S Architects & Associates

项目名称：Hanseong Baekje Museum
地点：88-20 Bangi-dong, Songpa-gu, Seoul, Korea
建筑师：YongMi Kim
项目团队：GilIlm Lee, YongJun Jang, YongIn Kim, JinHong Jo, GyuSaeng Hyun, DongJin Kim
合作商：Dongnam E&C
结构工程师：Hwan
设备工程师：Wowon M&E
机械工程师：Wowon M&E
电气工程师：Samoo TEC
景观建筑师：G.S.
承包商：Posco.Co.
甲方：Seoul City
用途：cultural and assembly facilities
用地面积：14,894.2m²
建筑面积：2,904.96m²
总楼面面积：19,423.06m²
建筑面积覆盖率：19.5%
总楼面面积比：18.23%
建筑规模：three stories below ground, two stories above ground
结构：steel framed reinforced concrete
竣工时间：2012
摄影师：courtesy of the architect – p.76~77
©JoonHwan Yoon(courtesy of the architect) – p.70~71, p.72, p.78~79, p.80, p.81, p.82, p.84
©JongOh Kim – p.94, p.95

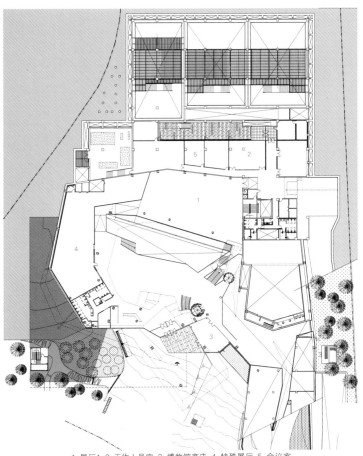
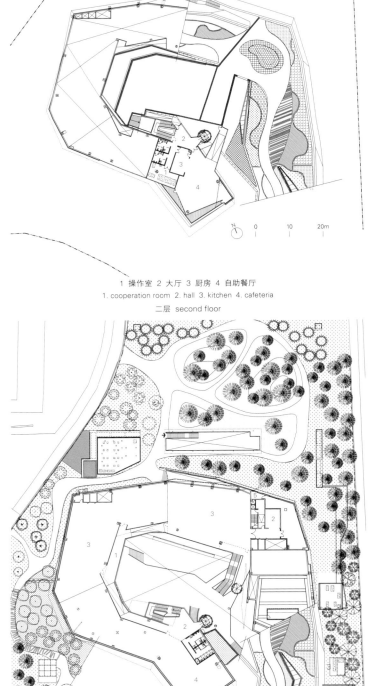

1 展厅1 2 工作人员室 3 博物馆商店 4 特殊展厅 5 会议室
1. exhibition-1 2. staff room 3. museum shop 4. special exhibition 5. meeting room

地下一层 first floor below ground

1 操作室 2 大厅 3 厨房 4 自助餐厅
1. cooperation room 2. hall 3. kitchen 4. cafeteria

二层 second floor

1 走廊 2 储藏室 3 图书馆 4 停车场 5 大厅 6 监控中心 7 工作人员室 8 货仓
1. corridor 2. storage 3. library 4. parking 5. hall 6. supervisory control center 7. staff room 8. warehouse

地下二层 second floor below ground

1 走廊 2 大厅 3 展厅2 4 展厅3 5 货仓
1. corridor 2. hall 3. exhibition-2 4. exhibition-3 5. warehouse

一层 first floor

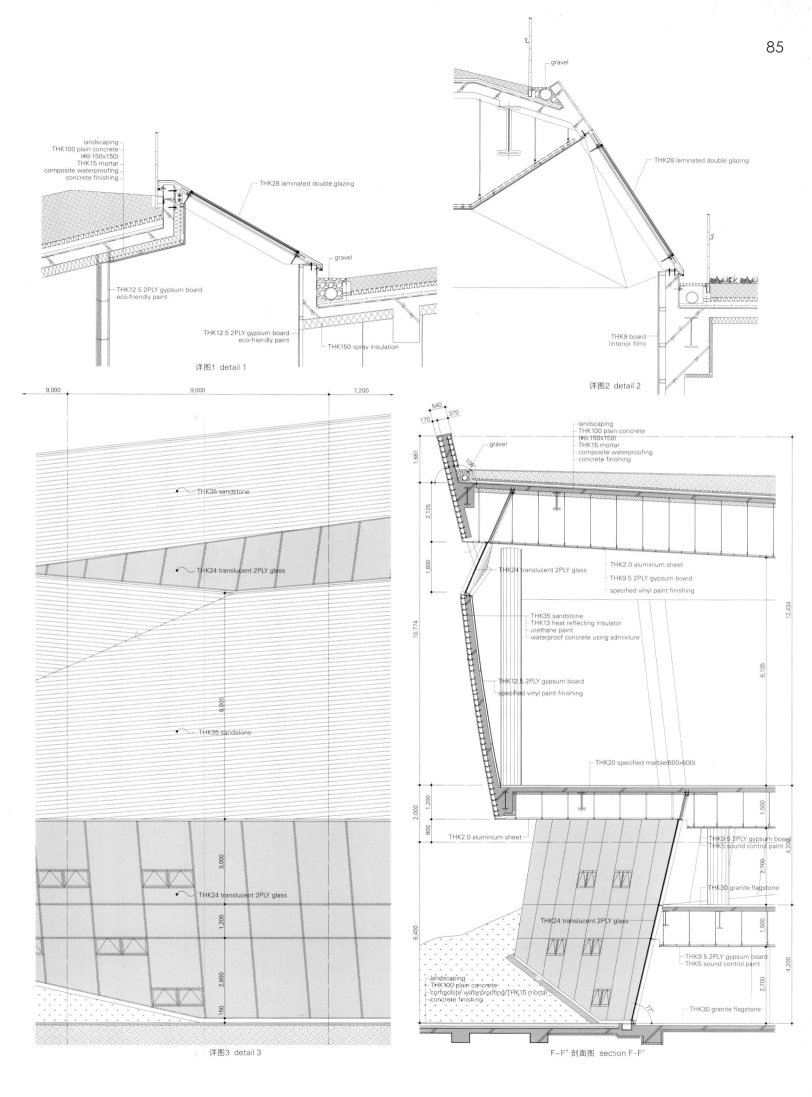

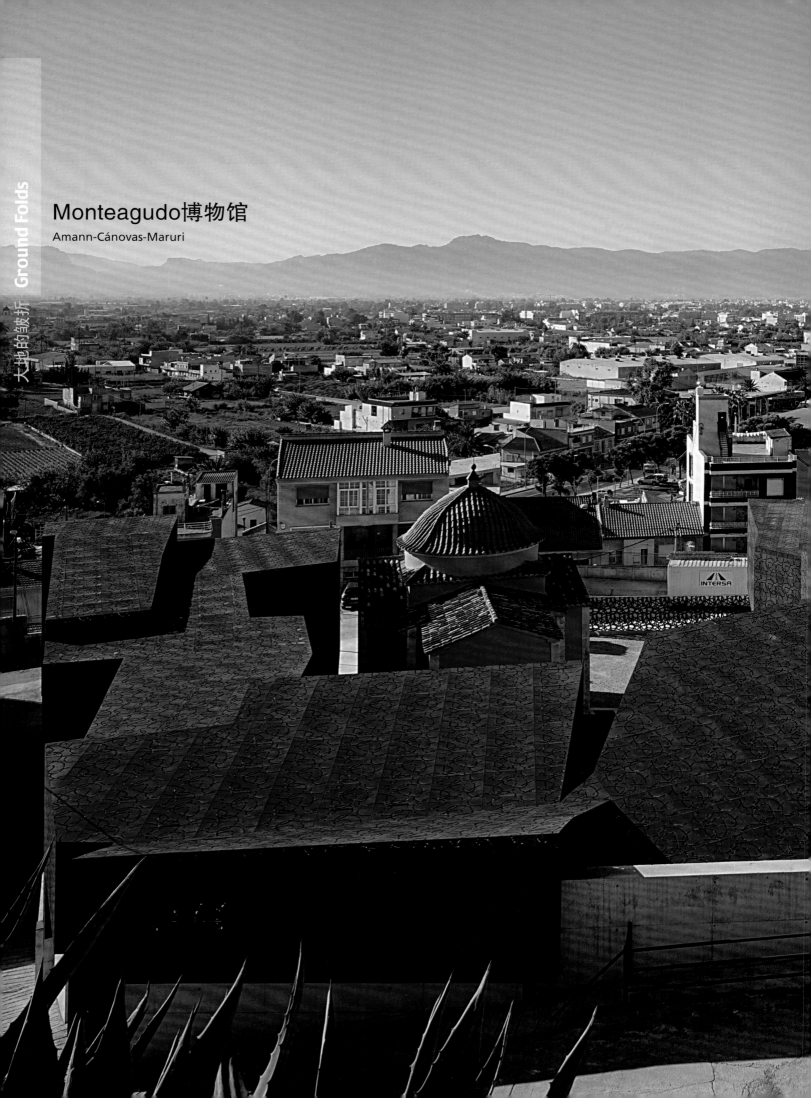

Monteagudo博物馆
Amann-Cánovas-Maruri

Ground Folds 大地的皱折

毫无疑问,这座建筑完全是依山而建的,位于Monteagudo山的南侧。它包含一个项目的首期工程,该项目旨在改善通往Monteagudo城堡的道路状况,想要通过修缮使Monteagudo城堡成为一个安全的旅游景点。

这片地域的历史可以追溯到史前,直到如今我们依然能看到从西班牙阿盖里克青铜时代开始,历经罗马文明和阿拉伯文明直至现在仍保留着的主要结构的遗迹。尤其是游客中心所在地,是保存状况很好的阿盖里克时期的村庄,也曾是罗马旧址。在同一地点还坐落着圣卡耶塔诺圣地,它赋予了这个地方某种特性。

建筑师的建筑提案在内容上与该环境的多重边界相适应,在忠于保护遗址功能的同时,从形式上和规模上也与现有地形相结合。建筑特别注重与山体相结合以及其从城堡方向所望的美景。

建筑既是一段上山的旅途,同时也是山体本身的一部分。作为上山的路线,它将道路融进斜坡,既解决了可到达性的问题,也使得该建筑部分与环境融为一体。

作为山体本身的一部分,建筑在颜色和形状上进行了很好的伪装,并且凭借建筑外表覆盖着的植物图案隐藏在周围风景之中。

一层设有对社区开放的公共呼叫中心。大型钢质网格墙体——有时可以滑动——和那些用混凝土粗糙地修建的房间都未经装饰,保护建筑的同时又与外部相连。这是一个处在阴影中的地方。

在上层,常设展厅和临时展厅都设在一处封闭而又受保护的区域内,该区域的唯一开口经过精心设计,是观看外面山谷和城堡的最佳视角。建筑也是一处观景点,窗子就像陈列柜一样,将室外的景色框起来供人欣赏。

建筑的一层有露石混凝土结构的围板和金属百叶窗,上层部分是金属结构,包含了向外的凸出部分并由带有加热防水密封层的多层板材进行封闭。最后,在上面再盖上一层穿孔的柯尔顿钢作为通风立面的最终层,起到气候监测的作用。

Monteagudo Museum

This building is, without a doubt, a parasite. The building is situated on the south side of the hill of Monteagudo. It constitutes the first phase of the project that demands to improve access to the castle of Monteagudo, by restoring it in order to get a safe place to visit. The slope of the mountain is a territory historically occupied since prehistory currently with remains of material structures that date to the Argaric world until today, through Roman and Arab civilizations. In particular, the site chosen for the Visitors Center is a Argaric village in a good state of preservation and a Roman site. In

tape motion

faceted wall motion

polyhedron motion

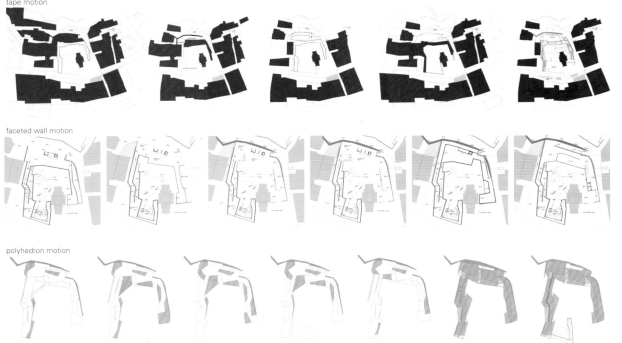

the same location, the shrine of San Cayetano is located, giving a certain character.

The building we have proposed tends to adapt to the multiple boundary from context, responding to the preservation of the remains and also consolidating the place from both formal and dimensional points of view, with special attention to integration into the hillside and its vision from the castle. The building is a trip and a parasite bloodsucking to the mountain. As a route, it resolves the accesses through the ramps that settle accessibility questions and the integration of the piece in the environment.

As a parasite, getting camouflaged in colors and shapes, carpeted with calligraphic plants skin that covers the building as a whole. The ground floor has a public calling as it sends out open to the neighbors. Its great steel lattice-worked walls, sometimes sliding, and those rooms rudely built with concrete show simply bare that provide shelter and connects to the outside. It is a place in the shade.

On the upper floor, the permanent and temporary showrooms are arranged in a closed and guarded place, which is only open in a careful way to the best views of the valley and the castle. Then, our building is also a viewpoint, a window changed into a showcase that get outdoors pieces framed to be grasped. About construction subjects, the building, on the ground floor, with exposed concrete structural screens and metal shutters. In the top section it runs with a metal structure that resolves the long juttings out and gets closed with a multiple sheets panel that is sealed with a heated waterproofing. Finally, the end is topped with a skin of perforated Corten steel, which works as a final layer in a trans-ventilated facade that takes up that old matter about climate as context. Amann-Cánovas-Maruri

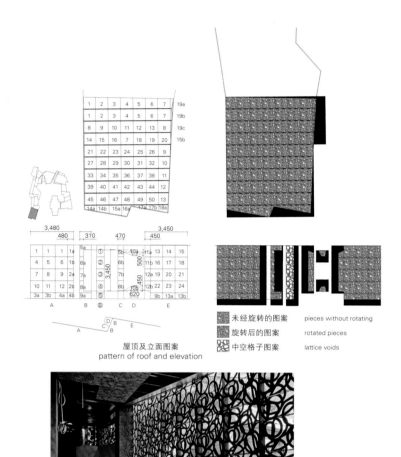

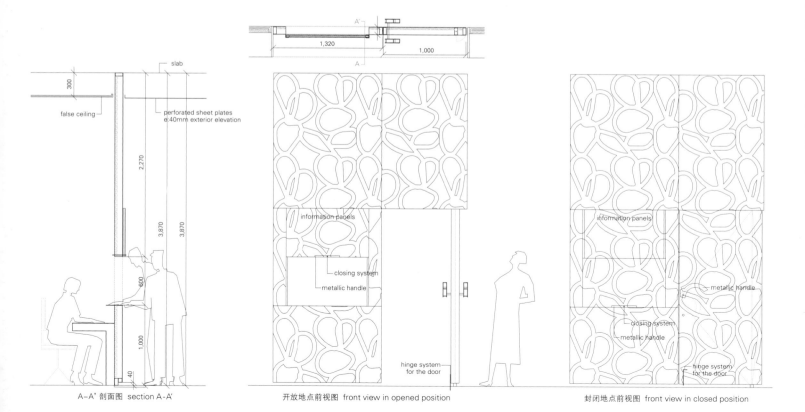

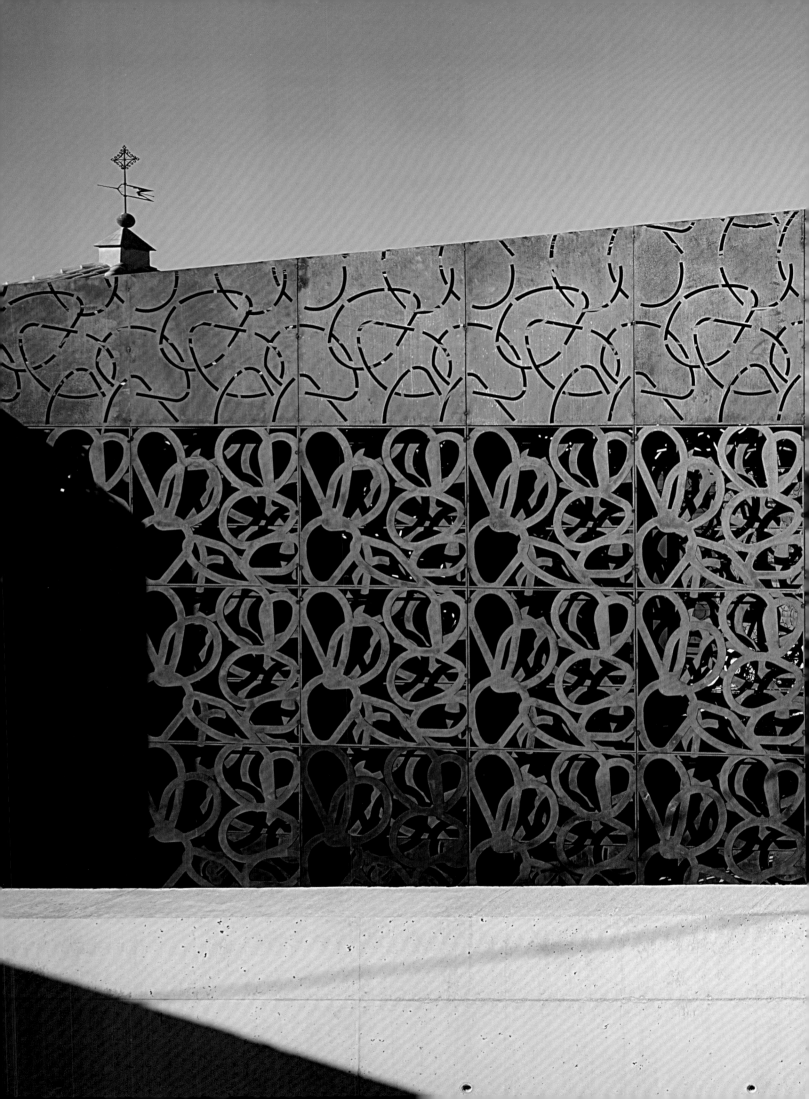

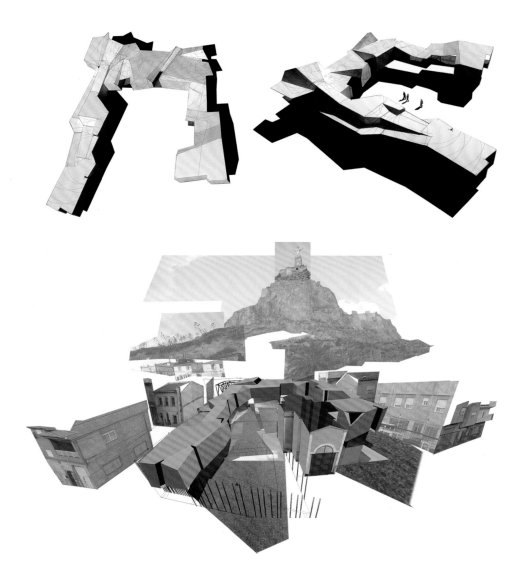
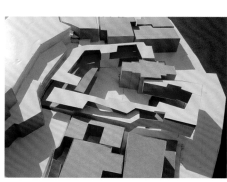
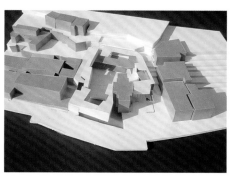

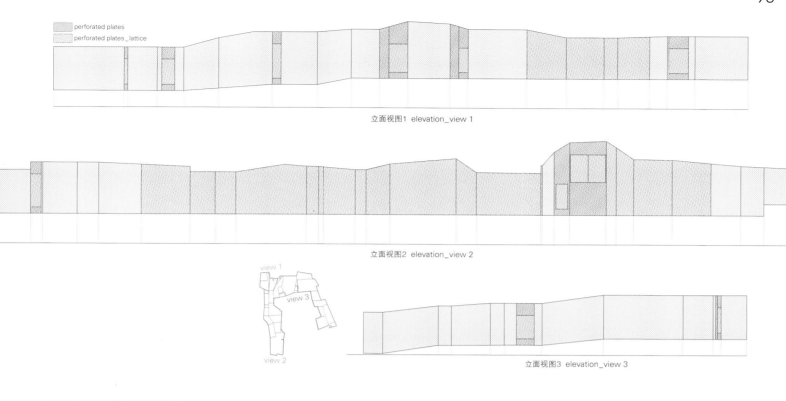

立面视图1 elevation_view 1

立面视图2 elevation_view 2

立面视图3 elevation_view 3

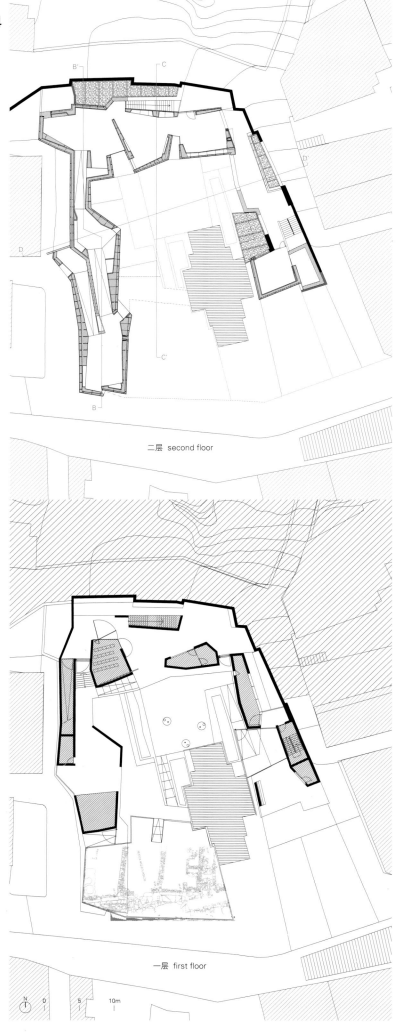

二层 second floor

一层 first floor

项目名称：Monteagudo Museum
地点：Plaza de San Cayetano, Monteagudo, Murcia
建筑师：Atxu Amann Alcocer, Andrés Cánovas Alcaraz,
Nicolás Maruri González de Mendoza
合作者：Javier Gutiérrez, Ana López, Patricia Lucas, María Mallo, Mónica Molero,
Carlos Ríos, Antonio Rodríguez
结构工程师：Ingeniería José Cerezo
机械及电气工程师：Ingeniería Condiciones Internas
质量检测员：Rafael Checa
甲方：Consorcio Turístico Murcia Cruce de Caminos
用地面积：1,264.73m²　建筑面积：1,058.54 m²
总楼面面积：391.75m²　施工时间：2008—2010
摄影师：©David Frutos (courtesy of the architect)

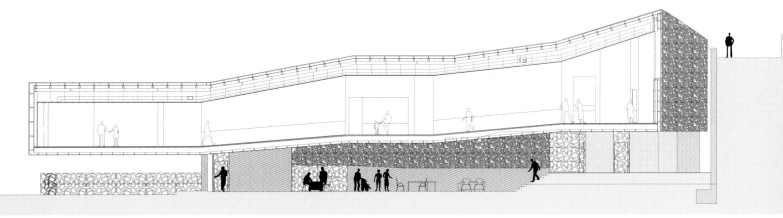

B-B' 剖面图　section B-B'

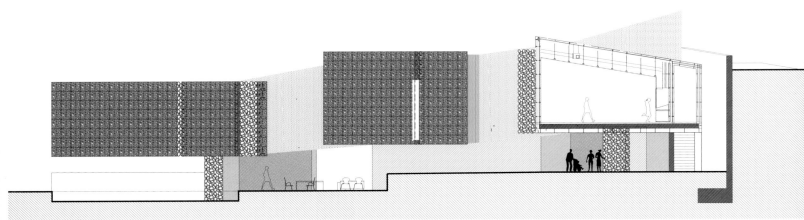

C-C' 剖面图　section C-C'

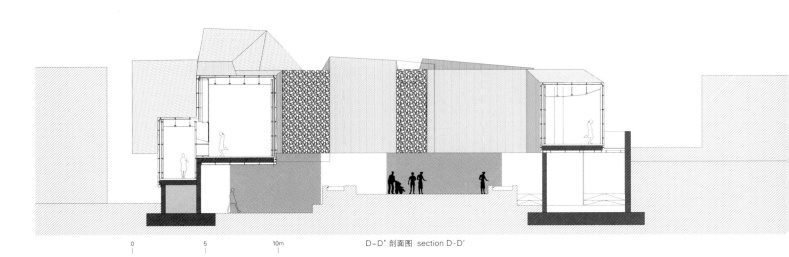

D-D' 剖面图　section D-D'

布鲁克林植物园游客中心
大地的皱折 | Ground Folds
Weiss/Manfredi

东立面1 east elevation1

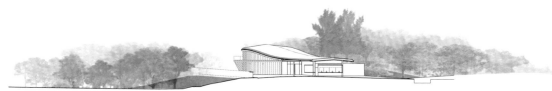
A-A' 剖面图 section A-A'

　　植物园其实是一个与众不同的博物馆——一个不断变化的、脆弱的收藏品。作为一处人工建造的"自然"环境,它更依赖人类建造的基础设施来繁荣发展。纽约市的布鲁克林植物园拥有大量不同的景观,这些景观经过布局形成连续的场景,如日式花园、樱桃园地、奥斯本庄园、眺望台、克兰福德玫瑰园。植物园在城市里就像绿洲一样,在视觉上通过抬高的坡台和树木将其本身和周围相隔开来。

　　为了激起人们对植物园里世界级收藏品的好奇心和兴趣,布鲁克林植物园游客中心提供了清晰的抵达点、方位图、公园和城市的交界处、文化点和培育区等。人们把该建筑设想成一块适合居住的地形,在城市和面积为210 437m² 的花园景观之间形成一个新的临界点。

　　场地位于华盛顿大街,建在将布鲁克林博物馆停车场和植物园分开的坡台中。游客中心提供了清楚的方位图和去公园主要景区(如日式花园和樱桃园地等)的路径。游客中心还有一条展廊、信息厅、向导室、礼品店、咖啡厅和活动室。

　　就像植物园本身一样,该建筑以电影的表达方式展现出来,人们永远无法从整体上看见它的全貌。游客中心曲折蜿蜒的结构是受植物园现存小路的启发形成的。从华盛顿大街进入游客中心的主要入口清晰可见。坡台顶部的第二条小径穿过游客中心,勾画出日式花园的整体景观,并通过一条有台阶的坡道向下通往主园林层。

　　游客中心展廊的弧形玻璃墙是建筑和景观之间的调节界面。玻璃的多孔表面过滤了阳光,把植物园的风景柔和地呈现出来。相比之下,游客中心的北面则完全隐藏在坡台之中。钢架结构的上层构造与这个具有弧度的平面相适应,赋予屋顶天篷起伏的形状。该建筑利用了土体和精选的多孔玻璃营造出高性能的建筑外围护结构,使热增量最小化,同时实现自然采光最大化。内部空间采用了地热交换系统来调节室温。其他的可持续策略则包括绿色屋顶、暴雨管理设施和用来浇灌一系列景观露台的雨水收集器。

　　游客中心的结构如同一个变色龙,从街道上的一座建筑转变为植物园的一个有序的风景。该中心重新定义了游客和公园之间的物理和哲学关系,并且引入了景观和结构、展览和运动之间的新关系。

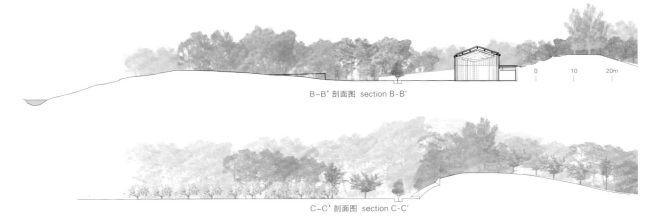

B-B' 剖面图　section B-B'

C-C' 剖面图　section C-C'

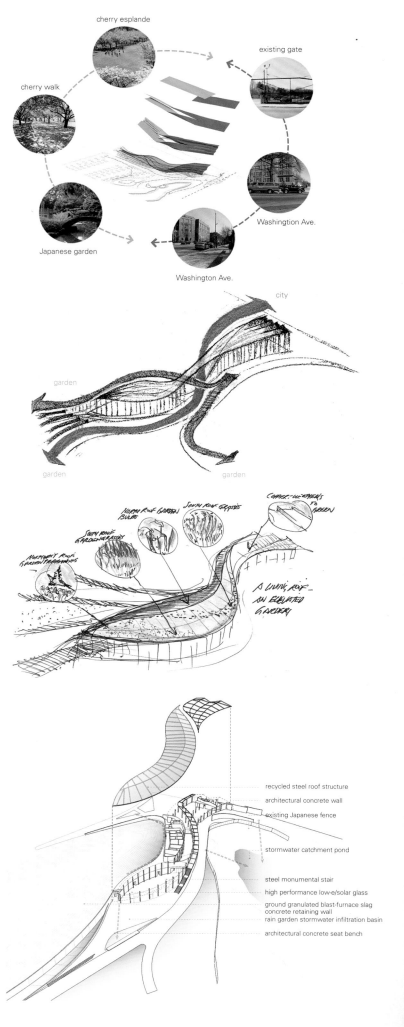

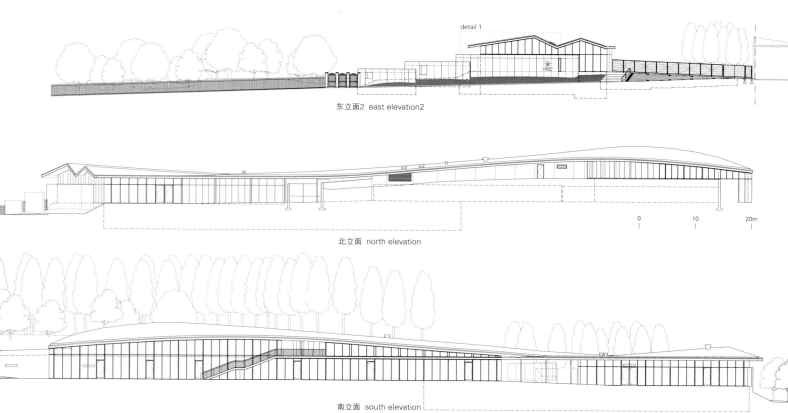

东立面2 east elevation2

北立面 north elevation

南立面 south elevation

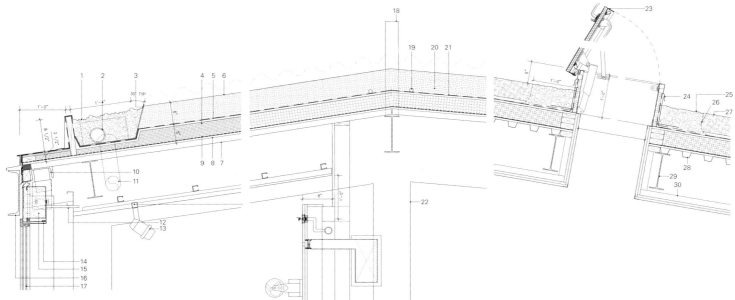

1. PVC waterproof membrane with crushed stone
2. 4" perforated PVC pipe
3. 4" perforated PVC pipe perforated SS basket(cont)
4. 4" rigid insulation
5. PVC waterproof insulation
6. green roof
7. 1-1/2" MTL deck
8. roof board
9. vaper barrier
10. MC6 W/ STL plate
11. rain leader
12. 6x18 AESS rigid frame
13. SS flashing, TYP
14. curtain wall head
15. batt insulation
16. 12" AESS channel fascia
17. 1" insulating glass
18. fall arrest system, 3-1/2" O.D. galvanized STL pier weld anchor to STL beam
19. drip irrigation
20. 6" green roof
21. PVC waterproof membrane
22. 6x10 AESS tube column
23. roof scuttle all edges to be fully sealed
24. CONT. fully sealed metal flashing edge all around
25. crushed stone
26. filter wrap
27. green roof planting bed
28. metal roof deck
29. STL structure
30. STL runner

详图1 detail 1

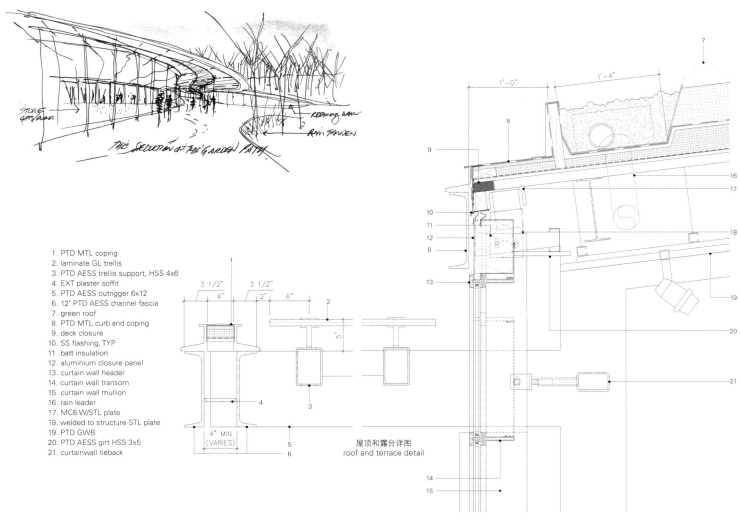

1. PTD MTL coping
2. laminate GL trellis
3. PTD AESS trellis support, HSS 4x6
4. EXT plaster soffit
5. PTD AESS outrigger 6x12
6. 12" PTD AESS channel fascia
7. green roof
8. PTD MTL curb and coping
9. deck closure
10. SS flashing, TYP
11. batt insulation
12. aluminium closure panel
13. curtain wall header
14. curtain wall transom
15. curtain wall mullion
16. rain leader
17. MC6 W/STL plate
18. welded to structure STL plate
19. PTD GWB
20. PTD AESS girt HSS 3x5
21. curtainwall tieback

屋顶和露台详图
roof and terrace detail

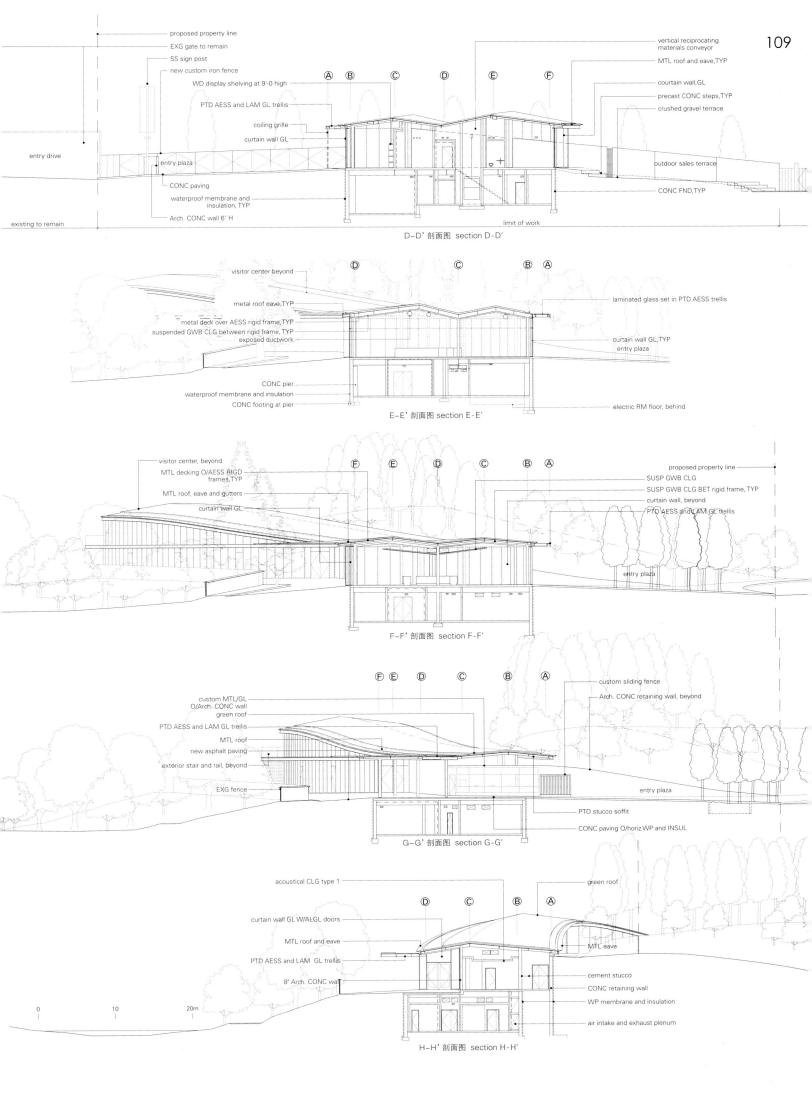

Brooklyn Botanic Garden Visitor Center

A botanic garden is an unusual kind of museum, a fragile collection constantly in flux. As a constructed "natural" environment, it is dependent on man-made infrastructures to thrive. New York City's Brooklyn Botanic Garden contains a wide variety of landscapes organized into discrete settings such as the Japanese Garden, the Cherry Esplanade, the Osborne Garden, the Overlook, and the Cranford Rose Garden. The Botanic Garden exists as an oasis in the city, visually separated from the neighborhood by elevated berms and trees.

To provoke curiosity and interest in its world-class collection, the Brooklyn Botanic Garden Visitor Center provides a legible point of arrival and orientation, an interface between garden and city, culture and cultivation. The building is conceived as an inhabitable topography that defines a new threshold between the city and the constructed landscapes of the fifty-two-acre garden.

Sited at Washington Avenue and within the berm that separates the Brooklyn Museum parking lot from the Botanic Garden, the Visitor Center provides clear orientation and access to the major garden precincts such as the Japanese Garden and the Cherry Esplanade. The Center includes an exhibition gallery, information lobby, orientation room, gift shop, café, and an event space.

Like the gardens themselves, the building is experienced cinematically and is never seen in its entirety. The serpentine form of the Visitor Center is generated by the garden's existing pathways. The primary entry to the building from Washington Avenue is visible from the street; a secondary route from the top of the berm slides through the Visitor Center, frames views of the Japanese Garden, and descends through a stepped ramp to the main level of the garden.

The curved glass walls of the Center's gallery are mediating surfaces between the building and the landscape. The fritted surfaces of the glass filter light and provide veiled views into the Garden. By contrast, the north side of the Center is inscribed into the berm. The steel-framed superstructure adjusts to the curved plan and gives shape to the undulating roof canopy. The building utilizes earth mass and spectrally selective fritted glass to achieve a high-performing building envelope, minimizing heat gain and maximizing natural illumination. A geothermal heat exchange system is used to heat and cool the interior spaces. Additional sustainable strategies include a green roof, storm water management, and rainwater collection that irrigate a series of landscaped terraces.

A chameleon-like structure, the Visitor Center transits from an architectural presence at the street into a structured landscape in the Botanic Garden. The Center redefines the physical and philosophical relationship between visitors and the garden, introducing new connections between landscape and structure, exhibition and movement. Weiss/Manfredi

项目名称：Brooklyn Botanic Garden Visitor Center
地点：Brooklyn, New York
建筑与场地设计：WEISS/MANFREDI Architecture/Landscape/Urbanism
设计合伙人：Michael A. Manfredi, Marion Weiss, FAIA
项目经理：Armando Petruccelli
项目建筑师：Hamilton Hadden, Justin Kwok, Michael Steiner
项目团队：Christopher Ballentine, Cheryl Baxter, Michael Blasberg, Paúl Duston-Muñoz
其他团队成员：Patrick Armacost, Jeremy Babel, Caroline Emerson, Eleonora Flammina, Kian Goh, Michael Harshman, Aaron Hollis, Hanul Kim, Hyoung-Gul Kook, Lee Lim, Jonathan Schwartz, Na Sun, Jie Tian, Yoonsun Yang
结构和土木工程顾问：Weidlinger Associates Consulting Engineers
信息技术工程顾问：Jaros, Baum & Bolles Consulting Engineers
地热和土力工程顾问：Langan Engineering and Environmental Services
景观：HM White Site Architecture
结构经理：The Liro Group
总承包商：E.W. Howell
甲方：Brooklyn Botanic Garden
用地面积：8,000m²
建筑面积：2,050m²
总楼面积：2,400m²
摄影师：©Albert Večerka/Esto(courtesy of the architect)

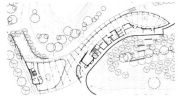

1. metal deck
2. gutter
3. PTD MTL coping and curb
4. PTD AESS channel
5. PTD aluminium fascia typical at CW
6. beam penetration
7. leader
8. radiant heat
9. trench drain
10. CONC MTL deck
11. STL structure W/fireproofing
12. CON pier
13. waterstop
14. filter wrap
15. footing drain
16. crushed stone
17. CONC footing
18. green roof
19. trellis glazing, LAM W/ceramic frit, SS fittings
20. CONC paving
21. acoustic ceiling
22. internal ELEC wiring
23. exit lighting fixture W/concealed backbox
24. exit manual pull station W/concealed back box
25. PVC waterproofing gutter W/washed stone
26. motorized shade
27. curtain wall GL W/ceramic frit
28. PTD AESS girt

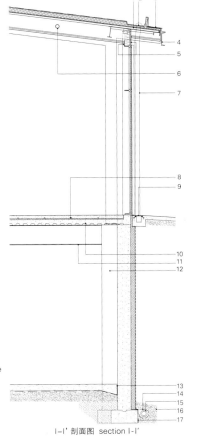

I-I' 剖面图 section I-I'

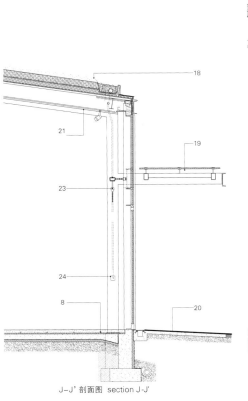

J-J' 剖面图 section J-J'　　K-K' 剖面图 section K-K'

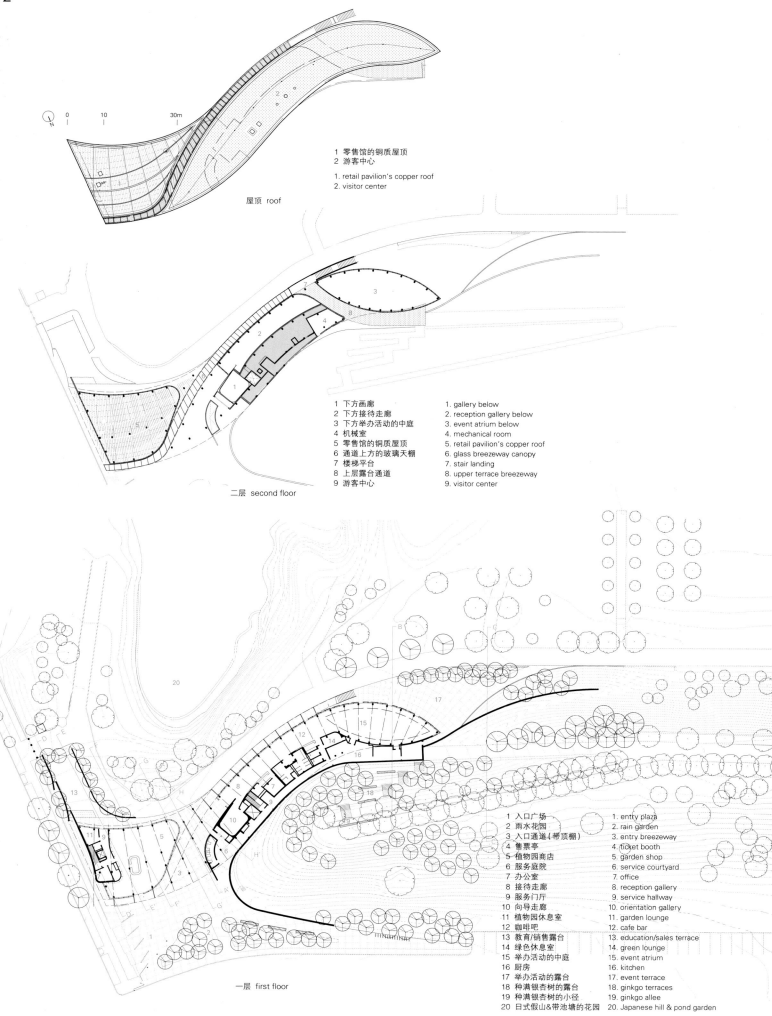

VanDusen植物园游客中心

Perkins+Will

VanDusen植物园游客中心坐落在温哥华市中心，它的设计初衷就是要从视觉角度和生态角度实现建筑和景观之间的和谐平衡。通过勘探和分析植物园的生态环境，设计团队能够将自然系统和人类系统融合，以恢复该场地的生物多样性和生态平衡。

设计团队精心地设计了建筑的绿色屋顶和周围的景观，使原生植物包含在其中，形成了一系列独特的生态区；植物覆盖的斜坡把屋顶和地面连接起来，吸引当地动物来使用；原始树木得到了精心的保护，并且促进了由湿地、雨林和溪水组成的生态系统的平衡。

受当地兰花外形的启发，游客中心在夯土和混凝土建成的弧形墙上建造了波形绿色花瓣状屋顶。它模仿自然系统，设计了集水、采光和储能系统以供需要时使用。1765m^2的游客中心位于植物园最显眼的东南角，改变了场地入口，从而提高了植物园的公众关注度，也强调了自然的重要性。坚厚的外墙让游客免受繁忙的街道的影响，透明的内墙使建筑向植物园敞开。中心设有咖啡馆、图书馆、志愿者服务设施、园内商店、办公室、可用作教室或租用地的弹性空间。

游客中心旨在超越LEED铂金认证等级，是加拿大第一座申请注册"生态建筑挑战"认证的建筑。"生态建筑挑战"是建筑领域最严格的可持续性衡量标准，由于它对项目有大量约束，如限制使用聚氯乙烯等含有有害物质的材料，因此全世界只有三个项目获得了完全认证。这一挑战也促使游客中心建筑取得革新式的成果——通过使用项目所在地可再生的地热钻孔、太阳能光伏和太阳能热水管等资源，实现了以年度为基准的纯零能耗。木材是最基本的建筑材料，能封存足够的碳以实现碳平衡；雨水过滤后用于建筑废水回收利用；所有废水都由温哥华第一台生物反应器现场进行处理，处理后排放到新型的渗流场和花园中。太阳能烟囱由一个开闭式玻璃圆孔和一个铝制穿孔散热器组成，能把太阳光转化成对流能源，从而促进天然通风系统工作。夏天的阳光照在深色的表面上，进一步加强了空气流通。太阳能烟囱处于中庭的中心，位于建筑物各种几何形辐射的正中心，因此，不论是在形式上还是在功能上，太阳能烟囱都凸显了可持续性的重要性。

预制屋顶结构都是由经森林管理委员会认证的花旗松组成，包含50多种预制屋顶面板，这些面板由100多种独特的弯曲胶合梁组成。预置屋顶面板中包含电路系统、灭火系统、视听系统等。由于要精确地制造出复杂的曲线造型，项目团队使用犀牛软件进行屋顶设计，通过快速原形三维立体印刷做出实体模型。工程文件是通过Revit软件运用三维技术产生的，这样能满足项目时间和预算都很紧张的底线。

VanDusen Botanic Garden Visitor Center

Located in the heart of Vancouver, the VanDusen Botanic Garden's new visitor center is designed to create a harmonious balance between architecture and landscape – from a visual and ecological perspective. Through mapping and analyzing the garden's ecology, the project team was able to integrate natural and human systems, restoring biodiversity and ecological balance to the site. The building's green roof and surrounding landscape were carefully designed to include native plants, forming a series of distinct ecological zones; a vegetated land ramp was included to connect the roof to the ground plane, encouraging use by local fauna; and old-growth trees were carefully preserved, facilitating an ecologically balanced system of wetlands, rain gardens and streams. Inspired by the organic forms of a native orchid, the visitor center is organized into undulating green roof "petals" floating above curving rammed earth and concrete walls. Mimicking natural systems, the building is designed to collect water, harvest sunlight, and store energy until needed. Located on the garden's prominent southeast corner, the 1,765m^2 visitor center transforms site's entrance to heighten public awareness of the garden and the im-

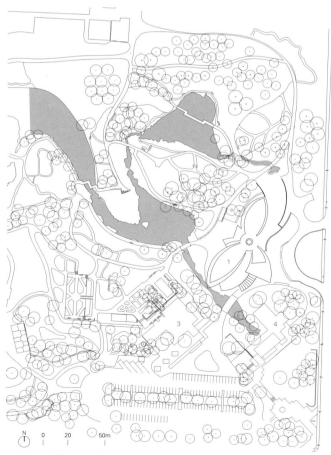

1 VanDusen植物园游客中心
2 橡树大街 3 行政办公楼 4 花卉馆
1. VanDusen Botanic Garden Visitor Center
2. Oak street   3. administration building   4. floral hall

西立面 west elevation

北立面 north elevation

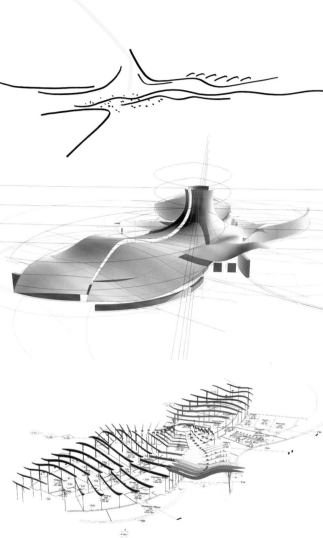
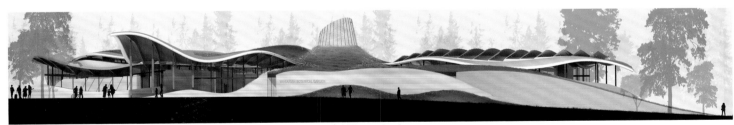

东立面 east elevation

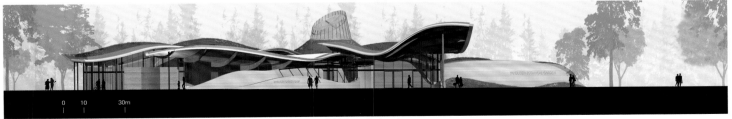

南立面 south elevation

portance of nature. With solid walls protecting visitors from busy street and transparent walls opening the building toward the garden, the visitor center houses a cafe, library, volunteer facilities, garden shop, offices, and flexible classroom/rental spaces.

Designed to exceed LEED Platinum, the visitor center is the first building in Canada to register for Living Building Challenge – the most stringent measurement of sustainability in the built environment. Placing enormous constraints on projects, such as restricting the use of Red List Materials like PVC, only three projects worldwide have earned full certification. The challenge also pushes buildings to achieve innovative results: the visitor center uses on-site, renewable sources – geothermal boreholes, solar photovoltaics, solar hot water tubes – to achieve net-zero energy on an annual basis. Wood is the primary building material, sequestering enough carbon to achieve carbon neutrality. Rainwater is filtered and used for the building's graywater requirements; 100% of blackwater is treated by an on-site bioreactor – the first of its kind in Vancouver – and released into a new feature percolation field and garden. Natural ventilation is assisted by a solar chimney, composed of an operable glazed oculus and a perforated aluminum heat sink, which converts the sun's rays to convection energy. Summer sun shines on darker surfaces to enhance ventilation further. Located in the center of the atrium, and exactly at the center of all the building's various radiating geometries, the solar chimney highlights the role of sustainability by form and function. Comprised entirely of FSC-certified Douglas fir, the panelized roof structure is composed of over 50 different pre-fabricated roof panels – made of over 100 unique curved glulam beams – that include electrical, sprinkler and audio-visual systems. With of the necessity to precisely develop the complex curve geometry, the team designed the roof with Rhino software and physical models generated with rapid prototype three-dimensional printing. Construction documents were developed three-dimensionally in Revit, enabling the project to meet a fast-tracked time and budget deadline.

Perkins+Will

1 入口大厅	1. arrival hall
2 中庭	2. atrium
3 办公室	3. office
4 室外商店	4. outdoor shop
5 餐饮服务中心	5. food service
6 志愿者服务中心	6. volunteer
7 服务中心	7. services
8 装卸区	8. loading bay
9 大礼堂	9. great hall
10 弹性空间	10. flex
11 教室	11. class room
12 图书馆	12. library
13 园内商店	13. garden shop
14 利文斯顿广场	14. Livingston plaza
15 利文斯顿湖边码头	15. Livingston lake dock

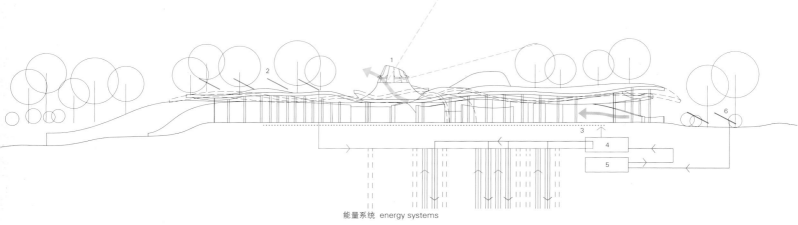

能量系统 energy systems

纯零能耗：现场使用的设施，可再生资源——地热钻孔、太阳能光伏设施和太阳能热水管的使用实现了以年度为基准的纯零能耗。
1 太阳能烟囱 2 太阳能热水 3 辐射板 4 热泵 5 电力换流器 6 光伏板
Net-zero energy: The facility uses on-site, renewable sources–geothermal boreholes, a solar photovoltaic array, solar hot water tubes–to achieve net-zero energy on an annual basis.
1. solar chimney 2. solar hot water 3. radiant slab 4. heat pump 5. electrical inverter 6. photovoltaic panels

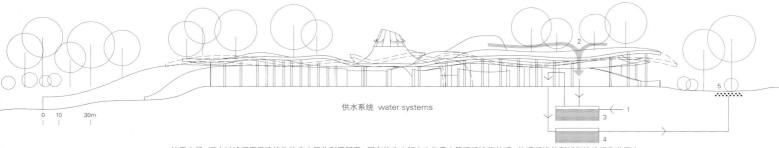

供水系统 water systems

纯零水耗：雨水过滤后用于建筑物的废水回收利用所需；所有的废水都由生物反应器现场进行处理，处理后排放到新型渗流场和花园中。
1 城市供水 2 顶板水收集 3 雨水收集和处理 4 废水处理/生物反应器 5 渗流场
Net-zero water: Rainwater is filtered and used for the building's greywater requirements; 100% of blackwater is treated by an on-site bioreactor and released into a new feature percolation field and garden.
1. city water supply 2. roofwater collection 3. rain water collection and treatment 4. blackwater treatment/ bioreactor 5. percolation field

项目名称：VanDusen Botanical Garden Visitor Center
地点：5151 Oak Street, Vancouver, BC, Canada
建筑师：Perkins+Will Canada Architects Co.
项目建筑师：Peter Busby, Jim Huffman
项目团队：Chessa Adsit-Morris, Aneta Chmiel, Paul Cowcher, Robert Drew, Benjamin Engle-Folchert, Robin Glover, Harley Grusko, Jacqueline Ho, Ellen Lee, Matthieu Lemay, Penny Martyn, Max Richter, Sören Schou
结构工程师：Fast + Epp
机械工程师：Cobalt Engineering
电气工程师：Cobalt Engineering
土木工程师：R.F. Binnie & Associates
景观顾问：Sharp & Diamond Landscape Architecture Inc. with Cornelia Hahn Oberlander
总承包商：Ledcor Construction
甲方：John Ross, Project Manager, Planning and Operations – Facility Development, Vancouver Board of Parks and Recreation
用途：Botanical Garden Visitor Center
用地面积：17,000m²
建筑面积：1,765m²
总楼面面积：1,765m²
竣工时间：2011.9
摄影师：©Nic Lehoux(courtesy of the architect)

太阳能烟囱由一个开闭式的玻璃圆孔和一个铝制穿孔散热器组成，能把太阳光转化成对流能源，因此它能促进天然通风系统工作。夏天的阳光照射在深色的表面上，进一步加强了空气的流通。
Natural ventilation is assisted by a solar chimney, composed of an operable glazed oculus and an aluminum heatsink, which converts the sun's rays to convection energy. Summer sun shines on darker surfaces to enhance ventilation further.

象征性纪念物

TEN Arquitectos

普埃布拉这座城市具有异常丰富的文化和传统,城市的每个角落都可以清晰地看到其悠久历史的印记,建筑师发现各种各样的建筑、不同的城市肌理、纷杂的声音交织在一起,构成当前的普埃布拉文化。在这个丰富多彩的文化马赛克中,建筑师能够发现在左卡洛(位于墨西哥城中心,是世界上最大的广场之一,译者注)的东北部有一个集文化、教育和娱乐为一体的建筑群,纪念碑将建于此地。

设计大赛的主题为"象征性纪念物",这往往让人以为是一个雕塑之类的物体,然而,这个方案新颖独特:充分利用当前的社会和建筑条件,创建一个起伏的空间结构。

项目通过创建新的开放公共空间,旨在激活和更好地利用场地,并且优化美丽的城市景观和视野。

纪念场地通过利用精心设计的空间所产生的空隙,来促进人与人之间的社会交往。

这些空隙包含了纪念场地的项目:一个圆形露天剧场、其下的一间多功能画廊、一个游乐场、荫蔽的开放空间和服务区域。

原有的地貌通过人工得到了再现,覆盖的木板层使公园看起来起伏波动。为了表达敬意,广场象征了墨西哥历史上最重要的历史性时刻之一,遍布广场的150棵树代表了普埃布拉战争150周年纪念日;广场植被的位置充分考虑了游客对纳凉舒适度的要求。灯柱的数量表明了作战的军队的数量,而该地的位置则表明最重要的战斗发生的地点。

这样,纪念场地成为一处不可思议的公共空间,它既为鸟瞰整座城市提供了一个很好的视角,又促进了社会和文化交流。

Emblematic Monument

The City of Puebla is a place with a vast richness in culture and traditions where every corner is a clear proof of its history made throughout the time; we can find a kaleidoscope of the buildings, textures and sounds that are all interwoven and create the current Poblana culture. Within this cultural mosaic, we can find on the northeast side of the Zócalo a complex of spaces destined for

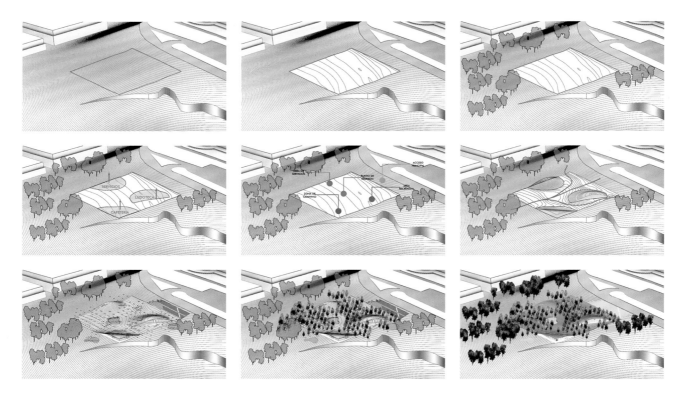

项目名称：Emblematic Monument of the 150 Anniversary of the Battle of Puebla
地点：City of Puebla, Puebla. México
建筑师：Enrique Norten
项目团队：Salvador Arroyo, Humberto Arreola, David Valencia, Ernesto Vázquez, Alejandra Téllez, Elsa Ponce, Miguel Rios, Erik Rico, Lenin Cruz, Liz Rebollo, Daniel Rios, Verónica Chávez, Sebastián Rodríguez, Isaac Uribe
承建方：GN Desarrollos SA de CV
分包商：INNDECO
甲方：Gobierno del Estado de Puebla
项目规划：urban space design with amphitheater, mixed-use gallery, playground, service area
用地面积：10,400m²
总建筑面积：11,580m²
设计时间：2011
竣工时间：2012
摄影师：©Luis Gordoa(courtesy of the architect)-p.132~133, p.134, p.135
©Pablo Crespo(courtesy of the architect)-p.129, p.133 top
©Patrick López Jaimes-p.128, p.130~131, p.133 bottom, p.136

culture, education and recreation; this site will host the memorial. While the competition called for an Emblematic Monument, often regarded as a sculptural object, this proposal creates a new project: an undulating spatial fabric made of social and architectural current conditions.

The project aims to activate and foment the use of the site by creating new open public spaces and reinforce the beautiful city views and landscape.

The memorial promotes social interaction through the use of the interstices generated by the deliberated design of the spaces. These gaps contain the program of the memorial: an amphitheater, a mixed-use gallery underneath the amphitheater, a playground, shaded open spaces and service area.

Artificial resemblance of the original topography is achieved, the overlapping wooden layer offers a park of undulating movements. Doing homage, the plaza symbolizes one of the most important historical moments in Mexican history. 150 trees were located throughout the square, representing the years that mark 150 anniversary of the Puebla Battle; the need of shade for the users' comfort determined the location of the vegetation. The light poles reflect the number of battalions that fought, while its location points to the sites where the most important battles occurred.

Thus, the monument becomes an unexpected public space, a viewpoint that looks over the city and promotes a social and cultural exchange. TEN Arquitectos

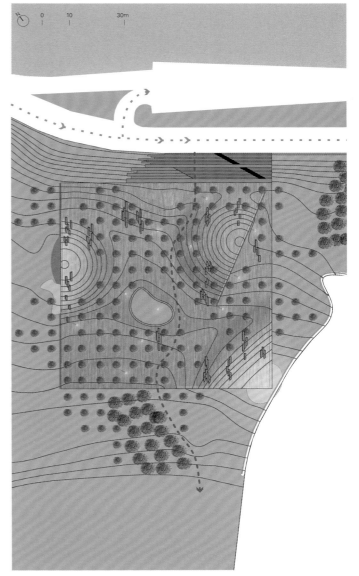

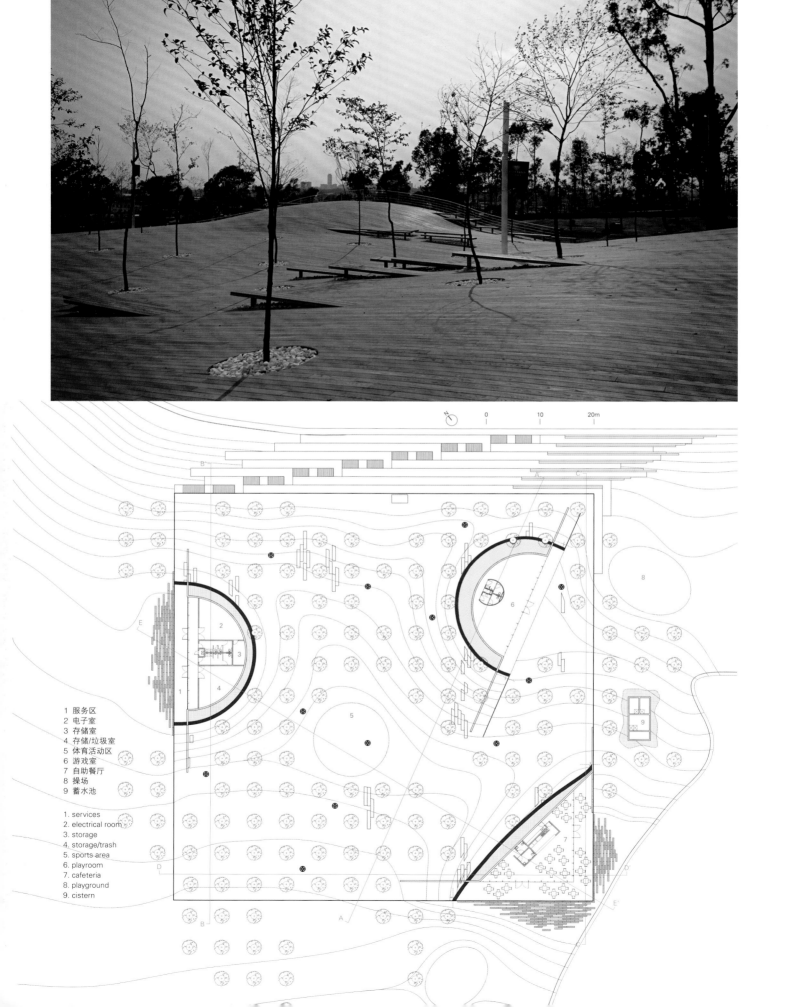

1 服务区
2 电子室
3 存储室
4 存储/垃圾室
5 体育活动区
6 游戏室
7 自助餐厅
8 操场
9 蓄水池

1. services
2. electrical room
3. storage
4. storage/trash
5. sports area
6. playroom
7. cafeteria
8. playground
9. cistern

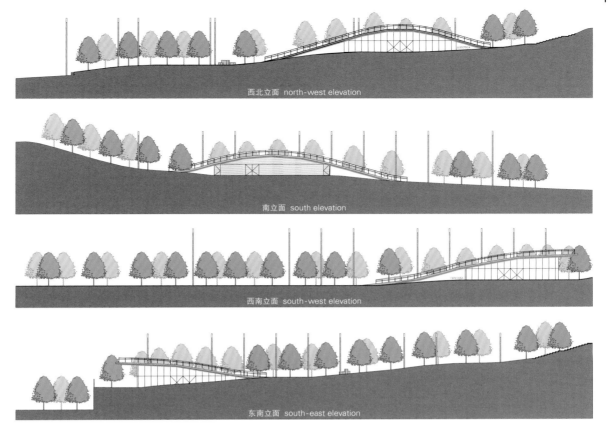

西北立面 north-west elevation

南立面 south elevation

西南立面 south-west elevation

东南立面 south-east elevation

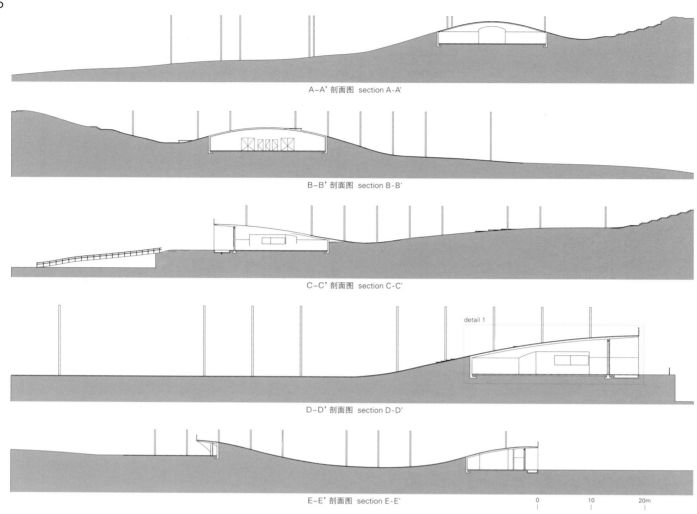

A-A' 剖面图　section A-A'

B-B' 剖面图　section B-B'

C-C' 剖面图　section C-C'

D-D' 剖面图　section D-D'

E-E' 剖面图　section E-E'

1. metalic frame made from galvanized 2x2" RHS@45cm
2. thermo-modified wooden deck
3. galvanized tube OC 10cmx6mm
4. anchoring system for the galvanized OC tube
5. concrete sleeper around perimeter
6. foundation block to hold OC
7. natural stone white gravel, 25mm grain size
8. metal fit for bench frame
9. metalic sleeper to hold galvanized tube
10. e=3mm metalic plate welded to galvanized RHS
11. cross-section galvanized 2x2" metalic RHS", anchored to concrete slab
12. natural ash exterior deck flooring. thermo-modified at 215°C planks, ranging from 1.80 to 3.60m long by 12cm wide, fixed with staples without showing any screws, leaving the surface completely smooth, SMA.
13. class 1 concrete(structural) f'c=250kg/cm² mixed with festegral powdered waterproofer

条凳详图 bench detail

1. 1 1/2" round base handrail number 30 anti-corrosive finish with white matte coating S.M.A.
2. handrail with round metalic base 1 1/2"x1/2" anti-corrosive with polyurethane Sylpyl 2010 finish color white 710
3. natural ash exterior deck flooring. thermo-modified at 215°C planks, ranging from 1.80 to 3.60m long by 12cm wide, fixed with staples without showing any screws, leaving the surface completely smooth, SMA.
4. metalic frame made from galvanized 2x2" RHS@45cm
5. steel 3/16" thick plate to hold metalic frame. anchored to concrete slab through metal 3/8" anchors.
6. 3"x1/4" angle with anti-corrosive finished with polyurethane Sylpyl 2010 color white 710, anchored to concrete slab.
7. dropper
8. class 1 concrete(structural) f'c=250kg/cm² mixed with Festegral powdered waterproofer
9. reinforcer made with #4 steel rods@20
10. anchors made of #4 steel rods
11. 30x25x0.95cm plate with anti-corrosive finished with glazed paint in white semi-matte S.M.A.
12. plain round screw 1.91cm
13. e=0.95cm plate with anti-corrosive finished with glazed paint in white semi-matte S.M.A.
14. CM-1 W8x31 column with inorganic zinc primer reinforced with epoxy and anticorrosive Sylpyl 3920 solvent-base, apply 2 coats of polyurethane Sylpyl 2010 color white 710
15. dividing CMU wall 12x20x40cm, concrete joints cement-sand, proportion 1:5, 1.0cm thick, reinforced with wire frame every 2 rows, with concrete steel rod No.3 reinforcers every 3m up until 5m height.
16. softwave 25 Hunter Douglas coating in perforated Aluzinc 0.5mm thick, post-painted in white, 0.28m wide by 4.50m long modules, placed vertically over the frame, with autoperforable visible fixation
17. RHS phrase framing a base de PTR Cal.14@1.20 max with first anti-corrosive and finished with white semi-matte coating paint S.M.A.
18. Type PT-01 6mm American oak plywood door. first class oak door frame 1 1/4"x1 1/4"
19. concrete slab f'c=200kg/cm² 5cm thick, in 1.20x1.20 modules with 6-6/8-8 wire netting
20. class 2 concrete slab(conventional) f'c=250kg/cm2. 12cm thick over compacted soil
21. 6x6-6/6 electro welded wire mesh
22. continuous concrete type 2 footing(conventional) f'c=250kg/cm²
23. concrete screed f'c=100kg/cm² 5cm thick
24. anchors made from #4 steel rods
25. MCA. Fondaline waterproof wire mesh model: Fondaline Plus S.M.A.
26. 1.5cm thick bituminous joint
27. precast concrete flooring in 0.30x1.50 modules granulated finish
28. dike made from health site material to 95%;
29. class 2 concrete retaining wall f'c=250kg/m² with festegral powdered waterproofer
30. reinforced concrete sleeper
31. healthy compacted earth filling

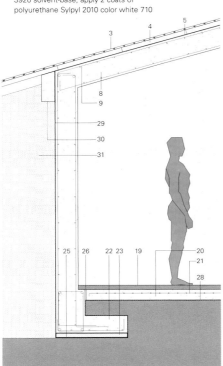

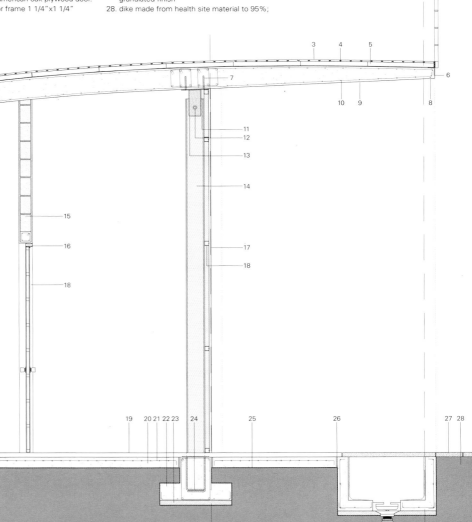

详图1 detail 1

都市改造与协调发展
Architecture and Recipro-City

Buso公寓／dmvA Architecten
模式化住宅／IND
青浦青少年活动中心／大舍建筑设计事务所
解码本土文化：相互调和的建筑设计／Nelson Mota

Apartments Buso/dmvA Architecten
Pattern Housing/IND
Qingpu Youth Center/Atelier Deshaus
Decoding the Vernacular: An Architecture of Reciprocity/Nelson Mota

  在过去的几十年里，人们对阿尔贝提的著名论断"房子如同一座小型城市，城市如同一座大房子"进行了不同的阐释，用来证明尝试在大规模建设中解决特色问题的建筑方法。第二次世界大战后，人种学方法受到青睐，用来指导建筑设计和城市规划。与杰出设计师阿尔贝提的观点一样，人种学方法认为人的尺度和社区的概念应该成为指导建筑师和城市规划者工作的主要参考因素。随后，福利国家的出现及其出现所带来的挑战为这一方法的诞生创造了条件。现今，发展中国家（经济体）经济的不平衡增长和世界经济体中前老牌核心国家的衰退使情况发生了变化，对于建筑与文化特色的关系产生了一系列新的挑战。

  荷兰建筑师阿尔多·范·艾克是人种学方法最积极的支持者，早在他作为CIAM（国际现代建筑协会）组织成员时就提倡这一方法，后来在成为建筑团体Team 10的核心人物之后仍然坚持提倡。范·艾克仔细研究了撒哈拉沙漠以南的社区和美国西南部普韦布洛人的当地建筑传统。这一经历使他提出了"温和的相互调节机制"的主张，认为要以此种机制来解决分歧，在创作中脱离"合适的尺寸"这个绝对的衡量标准。他的著名论断是：要达到合适的尺寸，不是回避这些典型的对立面，而是使它们相结合，如大小、多少、远近、统一性和多样性。

  这一主张似乎也是一些现代建筑方法关注的核心。尽管一些现代建筑方法兴起于迥异的文化背景，但是在重新评价本土文化特征和参与创建成功融合"双生现象"的景观进而唤回"自然"的自发性方面表现出了同样的兴趣。因此，现今，范·艾克所拥护的"介于中间"的概念再次盛行，成为克服日益增长的不均衡性的参考依据，从而做到在运动的状态中保持和谐。

Variations on Alberti's famous dictum "A house is like a small city and the city is like a big house". have been enunciated over the last decades to support architectural approaches that attempt to tackle identity issues in mass construction. In the aftermath of World War II an ethnological approach was favored as means to deliver an outcome that considered, as Alberti did, the human scale and the notion of community as key references for the work of the architect and urban planner. Then, the emergence of the welfare state and the challenges it brought about created the backdrop for this approach. Today, it is the unbalanced growth of the developing economies and the decline of the former core of the economic world system that shape the scenario where new challenges on the relation between architecture and cultural identity emerge.

Dutch architect Aldo van Eyck was one the most active supporters of that ethnological approach, advocating it both as member of CIAM and later of Team 10's inner circle. Aldo van Eyck carefully studied vernacular building traditions of both sub-Saharan communities and the Pueblos of the American southwest. This experience provided him an acknowledgment of the "mild gears of reciprocity", where polarities were tackled in such a way as to create something that goes beyond an absolute measure of "right-size". He famously declared that accomplishing the right-size is not avoiding but combining typical binary oppositions such as large and small, few and many, near and far, unity and diversity.

This also seems to be the core concern of some contemporary architectural approaches, which despite springing from disparate cultural backgrounds share the same interest in reassessing the qualities of the vernacular and contributing for the creation of a landscape where those "twin-phenomena" can be successfully combined, thus bringing back the spontaneity of the "natural". Hence, today, the notion of "the in-between", championed by van Eyck, is again gaining currency as a reference to overcome growing inequalities, and thus to dwell harmonically in motion.

阿姆斯特丹孤儿院，由阿尔多•范•艾克设计，1955—1960年
Orphanage in Amsterdam, by Aldo van Eyck, 1955~1960

照片提供：©Diego Terna

解码本土文化：相互调和的建筑设计
时代的融合：赞美自发性

"千篇一律"这一概念成了萦绕在当代建筑实践和话语中的挥之不去的标准，究其原因可知，战后经历了大规模的住房建设，使两次战争之间的一些现代建筑运动设计原则成为尽快、成本尽可能低廉地建造最急需的住房的工具。因此，尤其在欧洲和北美，标准化和大规模生产成为现代性的常见符号。现代主义城市规划原则所提供的理论装备，比如，1941年勒•柯布西耶出版的《雅典宪章》受到了那些在所谓的发达国家寻求解决住房紧缺问题者的珍视。可以说，赫鲁晓夫支持并拥护前苏联制定政策规定使用预制混凝土板材建造房屋正是这一方法的缩影。20世纪50年代中期直至20世纪70年代所修建的数百万平方米的赫鲁晓夫式住房（人们如此戏称）就是对此最好的见证。这一体系不仅在前苏联加盟共和国范围内全面使用，还传播到古巴、智利等遥远的国度[1]。显然，这一方法已经超出纯粹的技术或学科专业问题，与政治议题产生了共鸣，比如粉碎了个人主义倾向，把工人阶级从劳动密集型的生产方式中解放了出来，不一而足。

如今在我们的社会中，个人领域已经越来越占据主导地位，例如，像定制、变化和自由选择这样的概念已经成为市场营销策略的核心理念。亨利•福特曾在他著名的T型汽车座右铭"你可以选择任何颜色，只要它是黑色的"（意思为：只要汽车是黑色的，所有顾客都能随心所欲地把车涂成各种颜色。）中使用钝策略来解释消费者的选择，如今，这一策略已经被"超前定制"策略取代。例如，最近法国汽车制造商雪铁龙通过电视广告"雪铁龙DS3 Twins"所传递的"超前定制"市场营销策略。驱向定制和个人选择也会影响建筑领域，一是通过受市场驱动的求异商品化宣传力量来影响，或受到所进行的加强培育、谋求变化的激进项目的影响。

因此，自发性这一概念因追求建筑品质而赢得建筑师的拥戴。然而，就当代建筑实例来说，自相矛盾的是，这种自发性与其说是自然的，不如说是人为发明的。实际上，许多建筑师、政治家和开发商，还有其他代理商，现在都热衷于发明自发性，提交一个对多样性进行瞬间模拟的方案，而真正的多样性往往要经过时间的洗礼，伴随着使用者或是特定的情况所带来的改变。因此，这种情况人为地造成了不合时宜的建筑设计，在这种设计中，时间因素被压缩成一瞬间，这一瞬间通常融合了大量的原动力和设计过程共同作用的结果。

和谐运动与永恒

然而，这一现象并不新鲜，也绝不是我们时代所特有的，而是反

Decoding the Vernacular: An Architecture of Reciprocity
Amalgamating Time: In Praise of Spontaneity

The notion of monotony hovers over contemporary architectural practice and discourse as a haunted spectrum. A great deal of responsibility for this is due to post-war experiences with mass housing, which instrumentalized some of modern movement's interwar principles to deliver the much needed houses as fast and low-cost as possible. Thus, especially in Europe and North America, standardization and mass production became common tokens of modernity. The theoretical apparatus provided for by modernist urban plan principles such as the Athens Charter, published in 1941 by Le Corbusier, were cherished by those engaged in solving the housing shortage in the so-called developed world. Arguably the epitome of this approach was the Soviet policy of building housing using prefabricated concrete panels, championed by Nikita Khrushchev. The millions of square meters of Krushcheby housing, as they were nicknamed, built from mid-1950s until the 1970s testify for this. This system was thoroughly used not only within the borders of the Soviet republics, but also exported to such distant places as Cuba or Chile[1]. Obviously this approach goes way beyond a mere technical or disciplinary issue, and also resonates with political agendas, such as the shattering of the individual and the emancipation of the working class from a labor intensive mode of production, to name but a few.

The realm of the individual has become increasingly dominant in our society, and notions such as customization, change and free choice are central in marketing strategies, for example. Henry Ford's blunt strategy to account for consumer choice in his famous T-Model motto "You can choose any color, so long it's black", has been nowadays substituted by an "ultra-customizable" strategy conveyed recently, for example, by the French automaker Citroën TV advert "Citroën DS3 Twins". This drive towards customization and individual choice also affects the architecture discipline, both through the agency of market-driven discourses on the commodification of difference, or by engaged militant program fostering the accommodation of change.

The notion of spontaneity is thus gaining currency as a much sought after architectural quality. In the case of contemporary examples, however, this spontaneity is paradoxically more invented than natural. In fact, many architects, politicians and developers, among other agents, are nowadays keen on inventing spontaneity, delivering an instantaneous simulation of the diversity usually created through time with changes introduced by the users, or defined by specific circumstances. It thus deliberately creates an architectural anachronism, where the time-factor is compressed to one moment that amalgamates what usually is the outcome of a great deal of agents and processes.

Harmony in Motion and Timelessness

This phenomenon, however, is neither new and certainly not exclusive of our time; it is a recurring trend. Over the last six decades,

复出现的趋势。例如，近六十年来，我们发现许多建筑师都在为寻求更加人道地对待人类栖息地的方法而不断奋斗，努力寻找去掉人与人之间隔膜的建筑形式，这种隔膜主要是战后铺天盖地的大规模住房建设政策造成的。我认为，在这些立场中渗透着一种普遍的趋势：一种会把本土的东西作为自然的人文主义符号的驱动力。

例如，富有影响力的荷兰建筑师阿尔多·范·艾克亲自前往北非五国的原住民生活区，探访多贡人居住的位于撒哈拉沙漠以南的村庄，或去研究美国西南部的普韦布洛人，把自己的所见所闻用来研究这些居民定居点怎样建立个人与社区之间的平衡关系，进而对所谓的建筑构造原理[2]进行论证。对范·艾克来说，这将是能够成功控制多样性的一门学科，从而避免人们所认为的战后在处理所谓的大规模建筑审美方面的失败，并能够克服设计单调乏味的缺憾。从根本上说，范·艾克认为，结构主义体系的建筑学解释方式深受人种学方法的启发，认为建筑师应该设计这样的居住类型：在被重复仿制的时候，个体单元不会失去其特色，要不也应该能够获得一种延伸的特色，这样，就实现了他的和谐运动理论。人们认为，阿尔多·范·艾克设计的阿姆斯特丹孤儿院（1955—1960年）就是这一方法的经典之作。除此之外，还有他最得意的两个弟子皮特·布洛姆和赫尔曼·赫茨伯格设计的作品[3]。

所谓的现代主义推动了通用的建筑文化，由此带来了建筑设计千篇一律的问题。为解决这一问题，在范·艾克提倡建筑构造原理将近二十年后，美国建筑师克里斯托弗·亚历山大提出了自己的解决办法，即著名的"建筑模式语言"理念，建造我们感觉如家一般舒适温暖的建筑物和城市，充满活力，也就是他所说的"无名品质"。这一品质能建造出赏心悦目的建筑物或城市。因此，亚历山大认为，建筑师应该通过识别植根于特定文化和特定区域而长期形成的模式来寻求一种永恒的建筑方式[4]。他因而追随范·艾克的建筑理念，支持范·艾克使用本土文化作为参考的论点和宇宙论，认为对建筑和城市规划来说，大自然是最重要的符号。为培养自发性和多样性，亚历山大呼吁整个社会都参与设计，人人都是设计师，而不是仅仅依靠建筑师或城市规划者；每个人都应该在塑造一个富有活力的环境方面担负起责任。然而，他认为，要在整体中（例如社区或城市）明确有力地强调部分（个体）需要具备一定的条件。因此，他非常赞同范·艾克的论点，作为回应，提出了结构主义的构想，并将这一过程比喻为花与种子。他认为："花之所以完整且其细胞几乎都是独立的，是因为遗传密码，是遗传密码指导着各个部分形成的过程，使它们成为一个整体。"[5]对亚历山大来说，这一遗传密码具体体现在"良好的模式"上，可以共享和利用

for example, we can find many architects engaged in fighting for a more humanist approach to the habitat, and searching for forms to challenge the problem of the alienation of the individual by the overwhelming dissemination of standardization in post-war mass-housing policies. There is, I would argue, a common tendency that pervades these positions: a drive to bring about vernacular references as tokens of that natural humanism.

The influential Dutch architect Aldo van Eyck, for example, travelled to see the Maghrebian Kasbahs, to visit the sub-Saharan villages of the Dogon people or to study the pueblos of the American southwest, and used his experience in surveying how these settlements created a balanced relation between the individual and the community, to argue in favor of a so-called configurative discipline.[2] To van Eyck, this would be a discipline able to successfully govern multiplicity, thus avoiding the post-war perceived failure in coping with the so-called aesthetics of the greatest number, and to overcome the menace of monotony. Basically, van Eyck argued in favour of an architectural interpretation of structuralist systems, inspired by ethnological approaches, where architects should invent dwelling types where the identity of the individual unit should not lose its character when repeated, should otherwise acquire an extended identity, thus accomplishing what he named a harmony in motion. Aldo van Eyck's Amsterdam orphanage(1955~1960), is rightly considered a canon of this approach, to which are usually added works designed by two of his most cherished disciples, Piet Blom and Herman Hertzberger.[3]

Almost two decades after van Eyck's plea for a configurative discipline, the American architect Christopher Alexander contributed with his own solution to solving the problems brought about by the alleged modernist drive towards a universal architectural culture. He famously presented his idea of "a pattern language" as a way to create buildings and cities where we feel at home, that are alive, that have what he called a "quality without a name", which is the quality that makes them beautiful and pleasant buildings or cities. Alexander then argued that architects should seek a timeless way of building by identifying long established patterns rooted in each specific culture and particular place[4]. He thus follows van Eyck in supporting his argument using the vernacular as reference and a cosmological argument where the character of Nature is the ultimate token for both architecture and urban planning. To foster spontaneity and variety, Alexander calls for the participation of the society as a whole to become designers, and not just architects or urban planners; everyone should be co-responsible in shaping a lively built environment. He argues, however, that one should provide conditions to articulate the parts (the individual,) with the whole (the community, or the city, for example.) He thus echoes van Eyck's argument delivering a structuralist formulation for this process with the analogy of the flower and the seed. He claims *"what makes a flower whole, at the same time that all its cells are more or less autonomous, is the genetic code, which guides*

这些来模式设计和建造富有活力的建筑结构。

解码本土性

说来也怪，20世纪60年代末，那时正值范·艾克20世纪60年代早期形成自己的建筑构造原理和亚历山大在20世纪70年代末出版他的《永恒的建造方式》这一时间跨度的中间点，两人在秘鲁首都利马有一次"以建筑为名的邂逅"。1968年，在由秘鲁政府发起、联合国赞助为一个高密度低层住房实验项目举办了一场PREVI–利马住房设计大赛中[6]，两人都递交了参赛作品。然而，两人的设计对竞赛设计要求中所描述的"可扩建的房子"做了完全不一样的解释。亚历山大在设计中以建筑的语言诠释了本土生活方式和物质文化，认为设计应该考虑不同民族的喜好，而范·艾克更偏爱的房子是"其设计便于进一步自由发展，而不会违背居住者的最佳利益"。[7]因此，虽然两位建筑师在认可本土文化的品质方面惺惺相惜，皆认可本土文化的自然多样性和繁荣兴旺的自发性，但是他们的建筑设计表达方式却截然不同。我可以说，在以往的设计中也发生过类似的事情。虽然采用不同的方式，但项目设计师们都参考了本土的物质文化优良品质，其设计作品都依从当代对多样性的兴趣和对单调性的恐惧。

Buso公寓住宅综合体位于比利时的梅赫伦市，由比利时建筑事务所dmvA Architecten设计。梅赫伦市区主要由工业建筑构成，Buso公寓就位于这样的城市中，是对一个废弃工厂的改造，其周边建筑物林立，但是设计师对Buso公寓住宅综合体所处街区的设计注重虚实结合，可谓匠心独运。实际上，整个住宅综合体只有一小部分通过一个半开放的风化钢板饰面的体量与环绕建筑的Auwegemvaart大街相连通，看起来就像一扇大门，穿过它，去探究建筑物内部的其余设计。从Auwegemvaart大街到达建筑群还有一条路径就是穿过一条狭窄的小巷。小巷一边是鳞次栉比的排屋，一边是与其毗邻的工业建筑。这一段过渡是整个设计中最棘手的部分，因此才有了在地面层修建空中街道的设计，体现鲜明的城市特点。设计师独具匠心地利用空中街道达到由公共空间向私人领域过渡的效果，因此为体验这一过程提供了多种选择，如同我们在一个经历漫长岁月自然发展而来的小村庄闲逛也有许多选择一样。

青浦青少年活动中心由大舍建筑设计事务所设计，是尝试在建筑物中利用自发定居点的特性的又一成功案例。在这样一个有街道、广场或只是一个小院子的集合空间里，人们可以充分享受多样化的空间

the process of the individual parts, and makes a whole of them".[5] For Alexander, this genetic code is embodied in the "good patterns", those that can be shared and used to design and build something that is alive.

Decoding the Vernacular

Curiously enough, van Eyck and Alexander had an "architectural encounter" in the Peruvian capital city, Lima, in the late 1960s, somewhere halfway through the time span between van Eyck's early 1960s formulation of the idea of configurative discipline, and Alexander's late 1970s publication of *The Timeless Way of Building*. Both of them contributed an entry for the PREVI-Lima housing competition, a low-rise high-density experimental project launched in 1968 by the Peruvian government and sponsored by the United Nations.[6] Their projects, however, showed a rather divergent account on the competition brief's call for an expandable house. Whereas Alexander delivered an architectural translation of the vernacular life style and material culture into a project that should be able to accommodate peoples' preferences, van Eyck preferred a house that was "designed so that further free development cannot work against the best interests of the occupants".[7] Hence, though both architects shared a common interest in recognizing the qualities of the vernacular, it's natural diversity and blooming spontaneity, they found different ways to give it architectural expression. The same happens, I would argue, in the projects featured ahead. Though in different ways, their authors acknowledge the qualities of vernacular material culture as reference to deliver an outcome that complies with contemporary interest in diversity and fear of monotony.

In the residential complex *Apartments Buso*, designed by the Belgian practice dmvA Architecten in the city of Mechelen, the occupation of the block reveals an intelligent play of solids and voids with the existing volumes and urban fabric, which was chiefly defined by industrial buildings. Only a fragment of the residential complex is revealed in the main street surrounding it, the Auwegemvaart, through a partially open weathering steel cladded volume, which somehow creates something like a portal to discover the remaining parts of the scheme in the interior of the block. The access from the Auwegemvaart can be also accomplished by a narrow alley standing in the transition from a sequence of row houses and the neighboring industrial building. This processional transition intensifies the most notorious feature of this project, the arrival at a kind of elevated street in the first floor level, with evident urban characters. This instance of a street in the air, articulates with ingenuity the transition from the public to the private realm, thus offering several options to experience this process, such as the ones we would expect to have in a small village, spontaneously developed through time.

In the *Qingpu Youth Center* designed by Atelier Deshaus, one can find another instance of a successful attempt to use the qualities of spontaneous settlements at the scale of a building. The mul-

注释：

1. On the use of precast concrete panels in mass housing in the USSR and Eastern Europe Adrian Forty, *Concrete and Culture: A Material History*, Reaktion Books, 2012, pp.149~159
2. Aldo van Eyck's essay on the configurative discipline was first published in the Dutch journal Forum in 1962. This text was later included in Joan Ockman's anthology of post-war influential texts on architecture; See Aldo van Eyck's "Steps Toward a Configurative Discipline," in Joan Ockman, *Architecture Culture: 1943~1968*, New York: Rizzoli, 1993, pp.348~360
3. See, for example, Piet Blom's Hengelo's Kasbah, 1972-1974 or Herman Hertzberger's Centraal Beheer, 1967~1972
4. See Christopher Alexander, *The Timeless Way of Building*, New York: Oxford University Press, 1979
5. Ibid., p.165.
6. A good account on this project and its transformations over the last 40 years can be seen in Huidobro et al., *Time Builds!*, Editorial Gustavo Gili, 2008
7. See Aldo van Eyck's project description in "Previ/Lima. Low Cost Housing Project," Architectural Design, Vol.4, 1970, pp. 187~205

和社交体验，进而提高了培养青少年社会交往能力的可能性。然而，往往有一条界线把公共领域与私人领域隔离开来，社会交往互动的可能性受到了限制。在青浦青少年活动中心，公共领域和私人领域之间的过渡方式是非常微妙而循序渐进的，从而模糊了典型的公共领域和私人领域的划分。这一策略使青浦青少年活动中心的使用者和参观者获得了一种强烈的空间体验，这是一个探索与嬉戏的空间。这一空间体验得益于设计师对建筑的不同组件之间巧妙、均衡的配置。实际上，在这个设计中，我们谈论的不应该是建筑，而应该是空间分布体系的理念，这也是阿尔多•范•艾克所提倡的理念。

IND设计的模式化住宅位于北非的西班牙属摩洛哥休达，主要基于"模式化住宅"的理念用建筑语言阐释附近的"贫民区"（棚户区），既要应对复杂的社会和物质环境，又要面对通常所有社会住房项目都会存在的预算紧张问题，设计师用模式化住宅重新阐释了本土传统，在设计中引入本土传统文化的优秀特质，避免一味的模仿。因此，整个项目与地形巧妙地相互融合，这种培养集体主义感的创造性解决方案使整体具有很好的空间特质，也为人们体验整个楼群及其与周边风景的关系提供了多种可能性。一个巨大的平台把建筑与地形连为一体，就像是一张棋盘，镶嵌在棋盘上的高楼与平台平衡而和谐，让人联想起20世纪60年代毯式建筑的传统：雄伟的建筑让人们分不清到底是建筑物还是小村庄/群落。我认为，在邻近的地中海贫民区所成就的一切就是毯式建筑的翻版，因此，虽然没有采用家长式作风的设计方法，设计师也成功地完成了最初的目标。

在这三个项目中，我们当代的建筑师成功演绎了阿尔多•范•艾克和克里斯托弗•亚历山大应对战后大规模建设所带来的挑战和对人文尺度的削弱而提出的设计理念和原则。今天又有越来越多的建筑师开始关注本土文化，以期得到设计应该怎样使建筑更加和谐、更加体现人文主义关怀的答案。在这些经验之中，有一个看似深得上述所探讨项目的设计师们的推崇，也与15世纪阿尔贝提所发表的宣言相符，"房子如同一座小型城市，城市如同一座大房子"。尽管所采取的方式不同，有时甚至是冲突矛盾的，但范•艾克和亚历山大都用人种学方法回应了阿尔贝提的著名格言。如今，这些项目告诉我们，也许是再次解码本土文化和重视温和的相互调节机制的时候了。Nelson Mota

tiple spatial and social experiences an individual can enjoy in collective spaces such as streets, squares or just a small courtyard enhance the possibilities to foster social interaction. However, there is usually a boundary that separates these spaces belonging to the public realm from the private realm, where the possibilities for social interaction are limited. In the *QingPu Youth Center*, the transition between these two realms happens in a very subtle and progressive way thus blurring typical divisions between public and private. This strategy conveys this building's users and visitors an intense spatial experience, a space of discovery and playfulness, provided for by a neatly configured balance between the different components of the program. In fact, in this case, instead of talking about a building one could better use the notion of a configurative system, as Aldo van Eyck would advocate.

IND's *Pattern Housing* built in the north-African Spanish exclave of Ceuta dwells on the idea of "pattern housing" to deliver an interpretation of the neighboring "favela"(shanty town). Having to deal with a complex social and material context, and with the usual tight budget given to social housing programs, the pattern solution reinterprets the vernacular tradition and translates its qualities into a scheme that, nevertheless, avoids mimicking its reference. Hence, the project's clever adaptation to the topography and the inventive solutions to foster collectivity succeed in delivering spatial qualities to the ensemble and several possibilities to experience it and its relation with the landscape. The balance between the checkerboard towers and the platform that negotiates the articulation of the building with the site topography creates an outcome that could be affiliated with the tradition of the 1960s mat-building typology; a mega-structure, where we struggle to identify whether it is a building or a little village. This is, I would argue, exactly what happens in the neighboring Mediterranean favela, and thus the initial goal of the designers was successfully accomplished without a paternalistic approach though.

These three projects are thus good translations to our time of both Aldo van Eyck's and Christopher Alexander's solutions to cope with the challenges of post-war mass-construction and shattering of the human scale. Today again a growing number of architects look to the vernacular to get some answers on how to deliver a more balanced and humanist approach to the habitat. And one of these lessons, the one seemingly praised by the designers of the projects discussed above, echoes Alberti's 15th century dictum that *"A house is like a small city and the city is like a big house"*. Though in very different and sometimes conflictive ways, both van Eyck and Alexander echoed this motto using an ethnological approach. Today, these projects show us that perhaps it's time again to decode the vernacular and pay attention to the mild gears of reciprocity. Nelson Mota

Buso公寓

dmvA Architecten

作为城市空间的小巷

Buso公寓坐落在19世纪的工业轴心区域,位于梅赫伦—鲁汶运河沿岸,其所在地最大的特点是位于建筑林立的街区的中间位置。

小巷是进入的通道。项目所在地包含一座大型工厂建筑,一方面有大量临时搭建的小型建筑物,另一方面,有一个与Auwegemvaart相连的住所。外部空间既有限又零碎。

城市设计概念的目标是为工厂建筑再次带来新鲜空气并增加空间。第一阶段,人们拆除临时搭建的建筑体量,恢复厂房的模样。第二个阶段,为了能形成一个大的半公共空间,Auwegemvaart 19花园被占用。

这个相当封闭的厂房有两层。设计中,不时出现的屋顶和楼层片段形成天井和一条内部街道。阳光从开口处进入建筑。楼梯把半公共的内部街道和作为城市肌理的小巷连接起来。

到达所有的阁楼住宅都要经过这个半公共空间。

为了凸出旧厂房这一原址的精神,新建厂房的外墙都装有柯尔顿钢面板。

庭院和内部街道的内墙以及建筑物顶部新增加的建筑体量,表面都饰以聚碳酸脂面板。所以,先前的工厂成为新住宅区,形成了一场旧与新、厚重与透明、粗糙与光滑、黑暗与光明之间的交锋。

Apartments Buso

Alleys as Urban Space

Buso is located in the 19th century industry-axis, alongside the canal Mechelen-Leuven. Characteristic to the site is its location in the middle of a building block.

Alleys provided the access. The site enclosed a big factory building, with a number of small-added building volumes on one hand, and a dwelling with access to the Auwegemvaart on the other hand. The exterior space was limited and fragmentary.

The objective of the urban concept was to bring the factory building air and space again. In phase one, tearing down the added volumes rehabilitated the factory building. In a second stage, the garden of Auwegemvaart 19 was expropriated in order to get one big semi-public space.

The pretty closed factory building contained two floors. Here and there breaking down roof and floors generated patios and an inner street. The openings allow sunlight to enter the building. Stairs connect the semi-public inner-street with the urban tissue of alleys.

All loft-dwellings are accessed by way of this semi-public space.

To emphasize the spirit of the old factory-site, the outside of the factory is clad with panels of corten-steel.

The inner walls of patios and inner street, as well as the added volumes on top of the building, are finished with polycarbonate panels. So the former factory became a new residential area which is old versus new, massive versus transparent, rough versus polished and dark versus light. dmvA Architecten

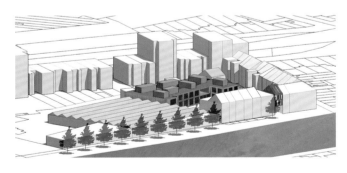

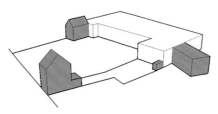

根据城市分析,拆除原有主体量的增建部分
removal of "additions" from the main existing volume according to the urban analysis

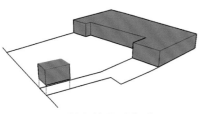

最大程度复原原有的厂房
maximal recovery of the existing factory building

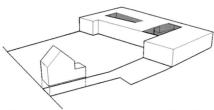

建造内部街道和天井,将自然光引入建筑内部,并建造横穿场地的小径
creation of internal street and patio to draw natural light inside the building and to create a pathway through the site

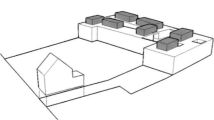

在屋顶上增建小型透明体量
addition of small transparant volumes on the roof

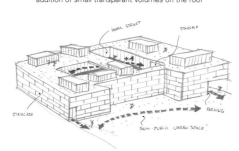

项目名称: Apartments Buso
地点: Mechelen, Belgium
建筑师: David Driesen, Tom Verschueren
项目团队: Astrid Geens, Christine Loos, Michaël De Roeck
结构工程师: Util
总承包商: Brebuild
建筑面积: 2,000m²
竣工时间: 2010
摄影师: ©Frederik Vercruysse(courtesy of the architect)

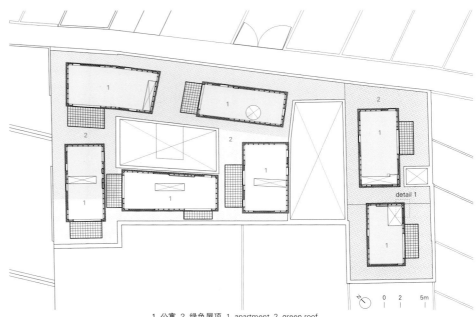

1 公寓 2 绿色屋顶　1. apartment　2. green roof
三层　third floor

1 公寓 2 内部街道　1. apartment　2. inner street
二层　second floor

1 停车场 2 天井 3 公寓 4 总储藏室 5 内庭院
1. parking 2. patio 3. apartment 4. general storage 5. inner courtyard
一层　first floor

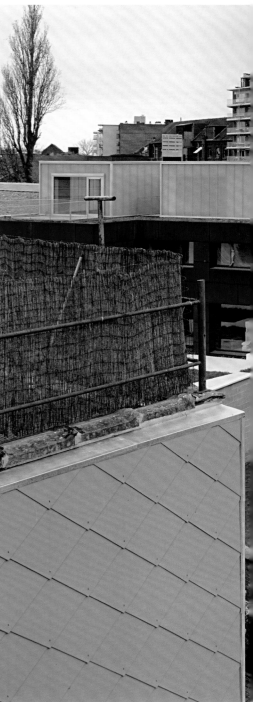

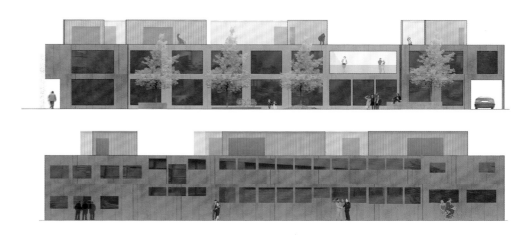

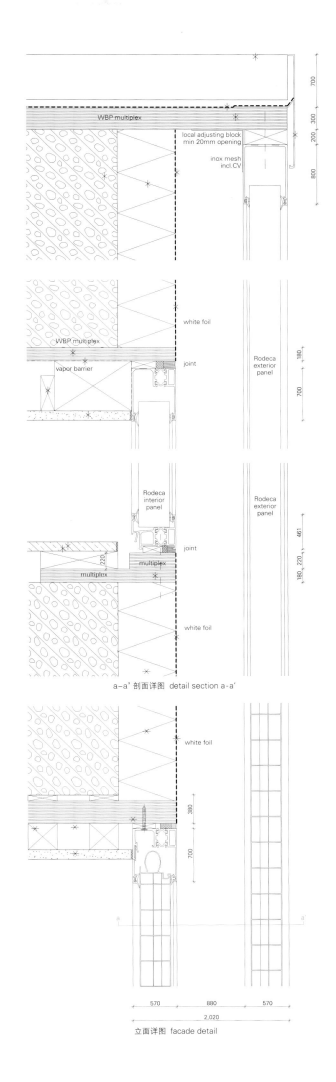

a-a' 剖面详图 detail section a-a'

立面详图 facade detail

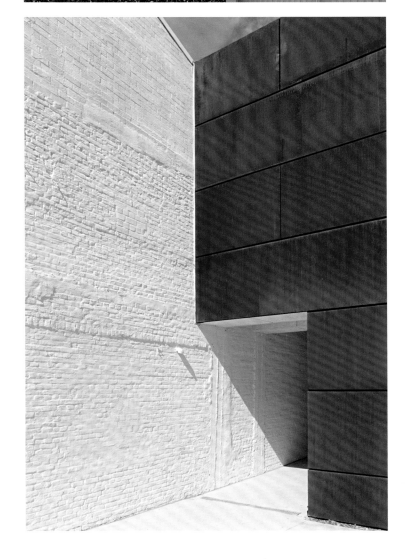

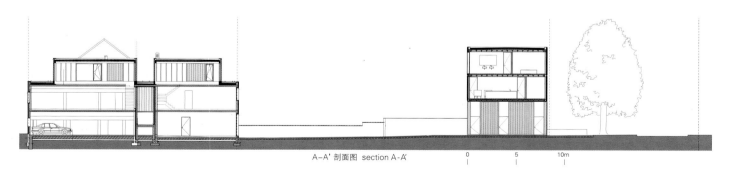

A-A' 剖面图 section A-A'

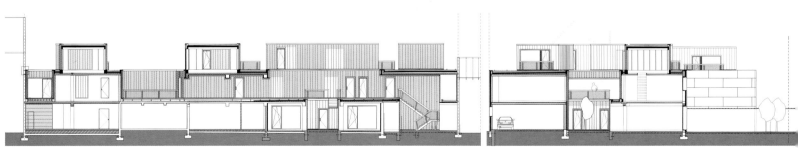

B-B' 剖面图 section B-B'　　　　　　　　　　　　C-C' 剖面图 section C-C'

模式化住宅
IND

先锋社会住房(VIVA)国际设计大赛由西班牙国家住房保障部组织，设计要求高度重视新建筑与城市环境的关系，宣传口号是"我们创造城市"。本社会住房项目是为了解决弱势群体的住房问题，同时与大多社会住房保障项目一样，预算紧张。建筑师决定通过利用项目位置的错综复杂性来应对紧张的预算。休达市位于非洲大陆的西北端，但它是欧洲西班牙的属地。如果你在地图上将休达市内本次竞赛所要设计的场地放大，就会发现不出1km就是摩洛哥的边界。从这个意义上来说，项目地点在欧洲和非洲的边界，然而一旦你到达那里，你不会觉得你是在一个西班牙的城市，也不会感觉是在摩洛哥的城市，它恰到好处地混合了不同的景观，这从某种程度上来说让人感到是独一无二的。在项目地点100m之内，放眼望去，你看见的也许是欧洲最大的非正式定居点，叫做Barriada Principe Alfonso。这个定居点有成百上千的自建房，居民大部分是阿拉伯移民。它规模不大、厚实、呈立方体状、迷宫似的街道模式，并且很少见那种将私人露台广泛用作公共空间的现象。进入此地，从附近清真寺传来的祈祷的声音不绝于耳，让你意识到这不是你所熟悉的位于城市周边地区的福利房景象。这儿是不同文化和环境强烈而美妙地融为一体的地方，但对非专业人士来说，这儿或许是两种不同的建城方式进行相当残酷较量的战场。最后，设计大赛的项目地点对社会住房项目来说也非比寻常，可以眺望优美的地中海风光。比赛要求设计方案在预算范围内有明显的可行性，并如大多社会住房设计方案一样，限制其类型为开放街区状(板条)。那么，他们的设计符合这些具体的条件吗？

因为设计大赛的第一阶段要求为950户住宅单元做一个城市计划，因此建筑师并没有把努力放在设计上，而是更多地集中在一个体系或一种模式上，从某种程度上说，是要继续探索克里斯托弗·亚历山大于1969年在利马的PREVI项目中所开创的空间体系，但他们只把他们的模式限制在建筑和空间元素上。建筑师把设计叫做"Vivienda de Patron"，意为"顾客喜欢的住房"，试图找到一种建筑类型来连接欧洲都市生活和周围的阿拉伯非正式定居点，并找到一个结合与合并两种不同的空间创立体系的模式。因此，这一模式既是一个迷宫，又是一个网格，每个单元的规模很小，但它与一个拥有停车场和储藏室等基础设施的大型基座合为一体，它还有私密露台供私人和公共使用，有开放的广场、亲切的街道和城市街角小店。项目的目标是为所有公民，无论其社会或经济背景如何，建造出舒适、漂亮的建筑。大多数公寓都有客厅角落，所有公寓楼之间都有狭长的视野，所使用的材料也是很少用于社会住房的材料，如大理石地板，还有许多像室内停车场和储藏室等实际好处。

你愿意住在那儿吗？在大赛中，建筑师多次放弃那些在形式上看起来非常吸引人的设计方案，因为对这个问题"你愿意住在那儿吗？"建筑师无法回答"是的"。因此，模式化住宅也是一个双关语，或者说是一个内部笑话，他们不是为穷人设计住宅，而是在设计一个顾客会喜欢住的居住空间。

项目名称：Pattern Housing
地点：Ceuta, Spain
建筑师：Felix Madrazo, Arman Akdogan
项目团队：Beatriz Zorzo Talavera, Angela Martinez Lago, Alvaro Novas Filgueira, Meritxell Rovira Pueyo, Elena Chevtchenko, Jose Manuel Franco, Michal Gdak, Melissa Vargas / Emin Balkis, Marta Koziol, Jorrit Sipkes, Inge Goudsmit, Harm Scholtens
场地方位：Alberto Weil Rus, Felix Madrazo, Manuel Perez Marin, Sanjay Prem
承包商：Acciona Infraestructuras
甲方：Spanish Ministry de Housing & SEPES
项目规划：170 social housing units, 1,500m² commercial, covered parking, storage
用地面积：7,153m²
建筑面积：21,292m²
总楼面比率：200%
施工时间：2009—2011
摄影师：©Fernando Alda

Pattern Housing

The international competition on Vanguard Social Housing (VIVA) organized by the Spanish Ministry of Housing placed a high emphasis on the relationship of the new architecture with the urban environment under the slogan "We Make City". Addressed to disadvantaged citizens and – as expected in most social housing projects – it had a tight budget. We decided to confront the tightness of the budget by taking advantage of the richness of the site complexity. Ceuta is in the northwestern tip of the African continent and yet it is a European enclave, if you zoom into the competition site within Ceuta you will find the border of Morocco within 1km. The site in that sense is in the frontier of Europe and Africa, yet once you are in the site you don't feel you are in a Spanish city, nor in a Moroccan, it is at the best a hybrid landscape and to an extent it feels unique in this regard. Within 100 meters the site faces perhaps the biggest informal settlement of Europe called Barriada Principe Alfonso, a settlement characterized by hundreds of self-built structures mostly inhabited by Arab immigrants characterized by their small scale, massiveness, cubic forms, laberynthic street patterns and intensive use of private terraces as public spaces are scarce. Once you are in the site the sounds of the call for prayer from nearby mosques makes you realize that this is not your classical peripheral social housing plot. It is an intense and beautiful merging ground of cultures and environments, or perhaps to an untrained eye a rather brutal battleground between two ways of making a city. Finally the competition site also plays an unusual role for social housing, with great views to the Mediterranean sea. The competition asked for a clearly feasible scheme within the budget and restricted its typology to open block (slabs) as expected in most social housing schemes. Was this the right answer for this specific condition?

Since the first phase competition of the asked for an urban plan for 950 units, we focused our efforts away from design and more towards a system or a pattern, in a way continuing the explorations of spatial systems initiated by Christopher Alexander in the PREVI project of Lima from 1969. But we restricted our patterns to architecture and spatial elements only. The design called "Vivienda de Patron" seeks to find an answer for a type of architecture that could be a link between the European urbanism and the Arab informal settlement in the surroundings, a pattern that combines and merges the spatial systems of both ways of making. Therefore it is both a labyrinth and a grid, it has a small scale in its units but it is also combined with a mega plinth for infrastructure such as parking and storage, it uses intimate terraces for private and common use with open squares, intimate streets and urban corner shops. The target is to build architecture generous enough to be attractive and comfortable for any citizen regardless of its social or economic background. Most of the apartments have corner living rooms, all have long views between blocks, materials seldom used for social housing such as marble flooring, and other practical benefits such as covered parking places and storage for all units.

And would you live there? Many times during the competition we discarded schemes that look formally attractive because we could not answer with a yes to the question: would you live there? Therefore Pattern Housing is also a pun, or an inside joke where instead of designing housing for the poor, we design a living space where the patron would like to live. IND

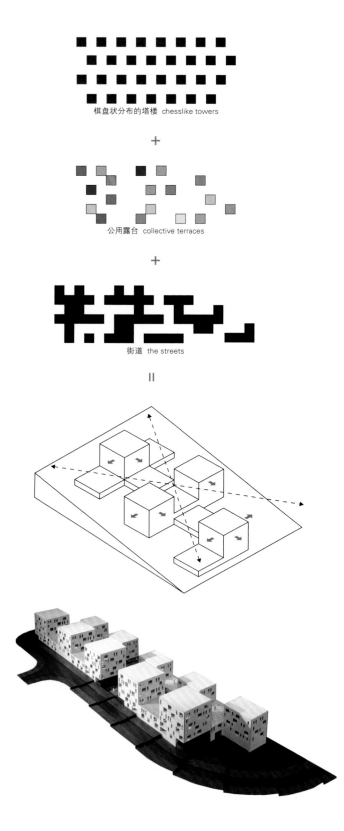

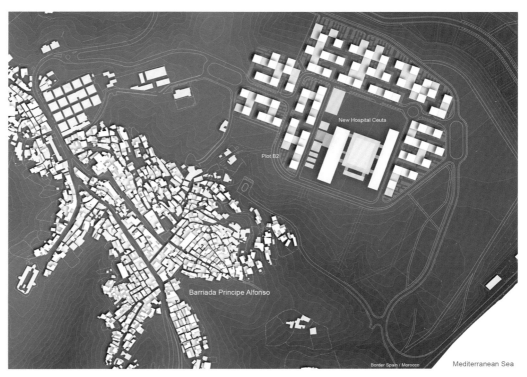

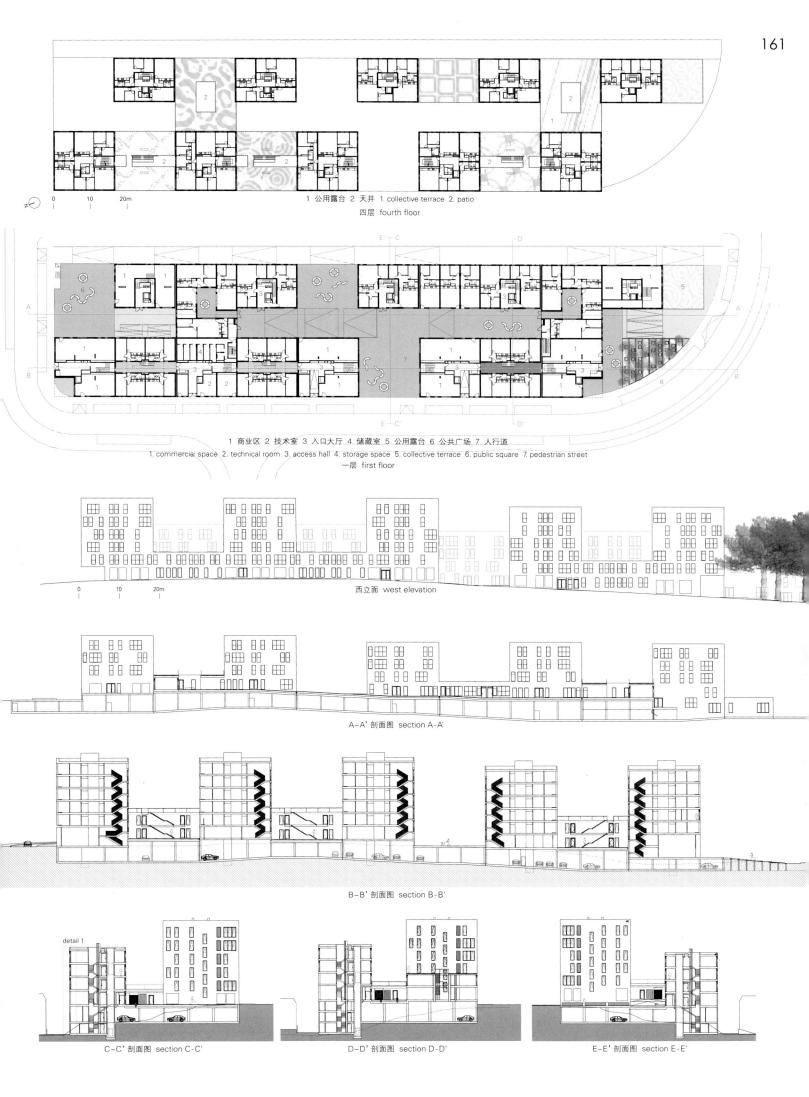

1 公用露台 2 天井 1. collective terrace 2. patio
四层 fourth floor

1 商业区 2 技术室 3 入口大厅 4 储藏室 5 公用露台 6 公共广场 7 人行道
1. commercial space 2. technical room 3. access hall 4. storage space 5. collective terrace 6. public square 7. pedestrian street
一层 first floor

西立面 west elevation

A-A' 剖面图 section A-A'

B-B' 剖面图 section B-B'

C-C' 剖面图 section C-C'   D-D' 剖面图 section D-D'   E-E' 剖面图 section E-E'

1. blunt grave, 5cm
2. geotextile sheet
3. extruded polystyrene EPS 4cm
4. water resistant sheets (bitumin, asphalt, geotextiles)
5. cement mortar, 2cm
6. lightweight concrete 1.5%, 10cm
7. asphalt vapor barrier
8. unidirectional slab, 25+5cm
9. gypsum plaster, 1,5cm
10. aluminium plate, 1cm
11. selfprotected membrane
12. expanded polystyrene, 2cm
13. concrete top slab
14. forated brick, 24x11, 5x7cm
15. monolayer coating
16. insulating mortar
17. ursa glasswool panel mur
18. opaque glass balustrade, H=160cm
19. aluminium flashing, 2cm
20. oxidated asphalt sheet
21. concrete riddle (carpentry support)
22. marble tiles
23. regulation mortar screed, 5cm
24. polyester sheet, 5cm
25. ursa XPS panel, 3cm
26. fire resistant plaster
27. prefab. concrete lattice panel, 12cm
28. foundation plate, 20cm
29. backfill layer, 20cm

详图1 detail 1

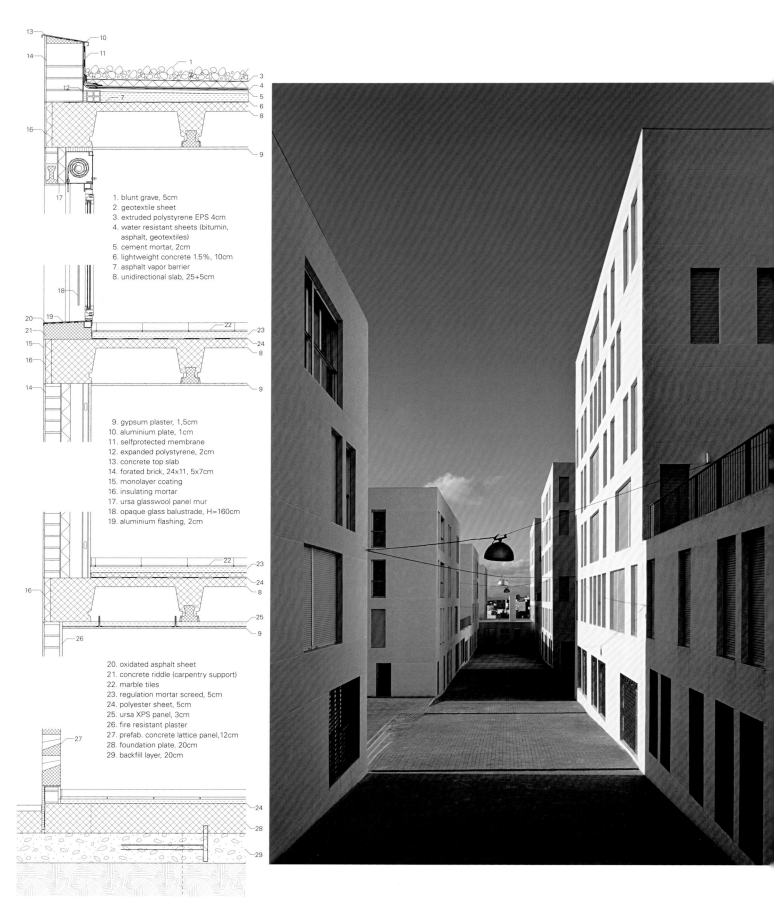

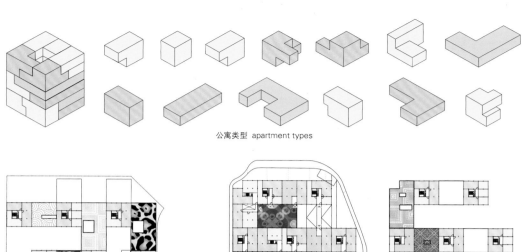
公寓类型 apartment types

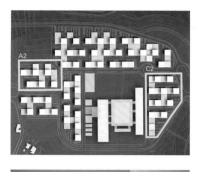

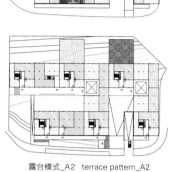

露台模式_A2 terrace pattern_A2

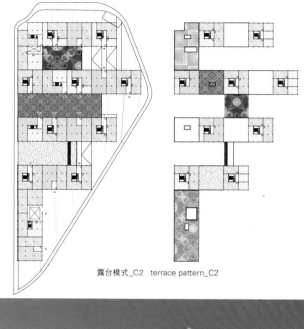

露台模式_C2 terrace pattern_C2

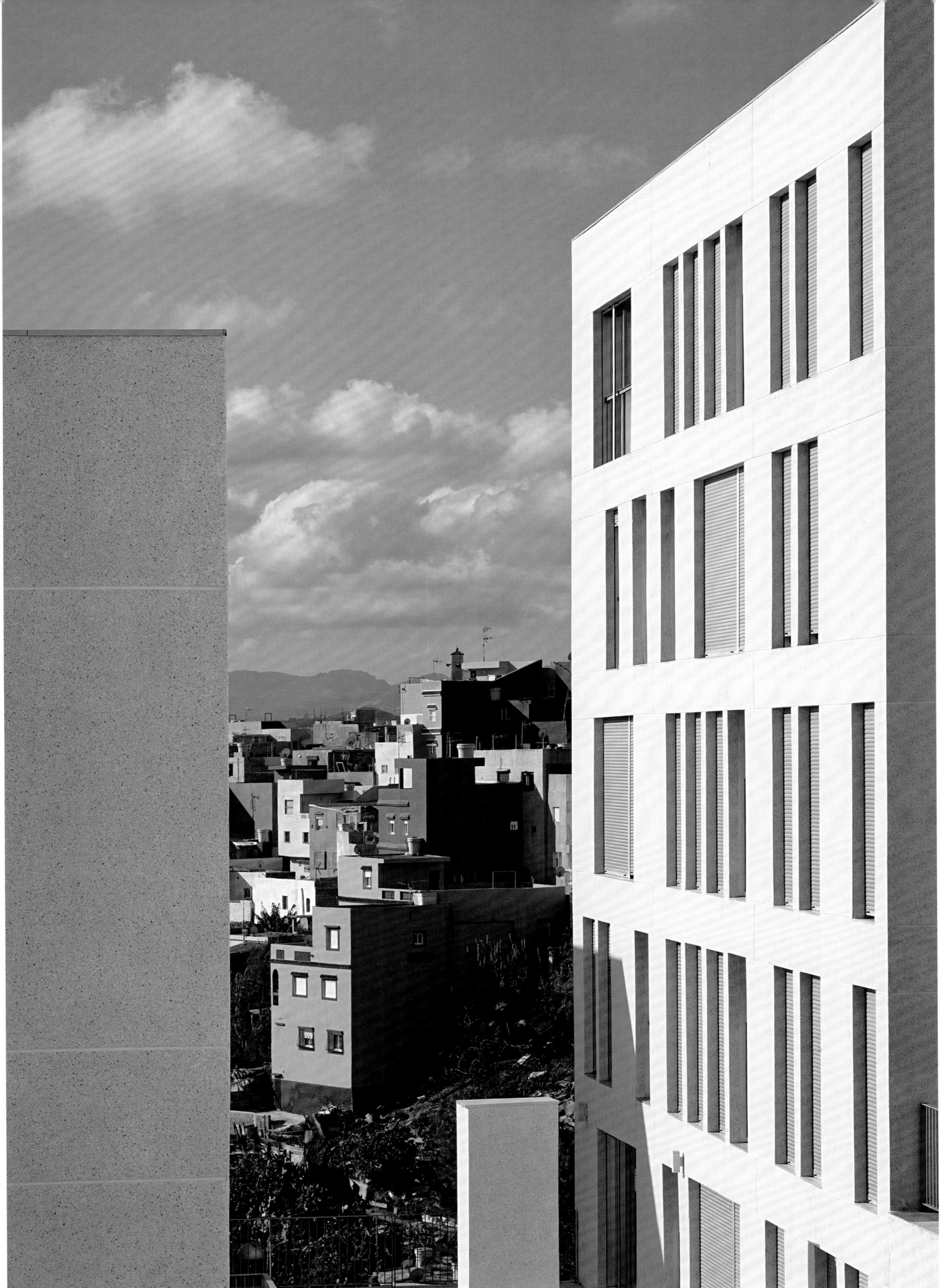

青浦青少年活动中心

Atelier Deshaus

青浦青少年活动中心位于上海市青浦区的东部新城。青浦新城相对老城而言，空间规模变大了，笔直的道路四通八达。因为统一而机械的规划和控制，建筑离道路的距离呈现出单调与疏离的布局。空间规模的变化源于交通模式的变化和人口的增长，建筑风格则受到现代城市文化的影响。

在青浦新城的主干道上，传统江南小镇富有人性化的小规模空间已基本消失了，但一旦进入次一等级的公共空间，人们仍可以将其与江南某些方面的相关记忆联系起来，如北菁园园林、行政学院南侧的小河、华青路沿岸的小河，还有尚未被拆迁的夏阳村及其鱼塘等，这一切都提示我们在大规模城市的压力下，仍然有构建人性化小规模公共空间的可能。

青浦青少年活动中心建筑根据具体的使用特点将不同的功能空间分解开来，转化为规模相对较小的单元，再利用庭院、广场、街巷等外部空间将其连为一起，从而成为一个建筑群落的聚合体。青少年在其间的活动——不同功能空间之间的连接、漫无目的地游荡以及随机的发现——就像在小城市里的活动一样，这是建筑师对郊区城市化过程中日益扩大的城市规模所做出的回应。建筑师希望在城市建筑规模已经扩大的前提下，仍能创建一个内在的人性化的小规模公共空间，重拾小规模的传统江南小镇的记忆。

在传统的江南居住空间中，建筑的外部空间和内部空间常常是同等对待的，甚至外部空间比内部空间更为重要。本案设计秉承了这样的传统，而符合人性化的、有趣的外部空间也极好地适应了青少年的性格和活动特点。

一座建筑，也可以是一座小小的城市。

Qingpu Youth Center

Qingpu Youth Center is located in eastern new town of Qingpu. Compared with the old town, the new town of Qingpu hugs a larger scale with straight roads of high availability. Due to unified and mechanical planning and controls, the distance between buildings and roads takes on the layout of platitude and alienation. Scale change derives from changes in traffic patterns and population growth while architectural style is affected by modern urban culture.

In main roads of new town of Qingpu, the personified small-scale space of traditional south river towns has basically disappeared. But once entering public space of lower level, you can still be connected with some memories of fields related to regions south of the Yangtze River, such as gardens of Beijing Park, the river south to the executive college, the river along Huaqing Road, Xiayang Village which has not been demolished yet and its ponds etc., which prompts us that under the pressure of large-scale city, there are still possibilities for building small-scale public space of humanization.

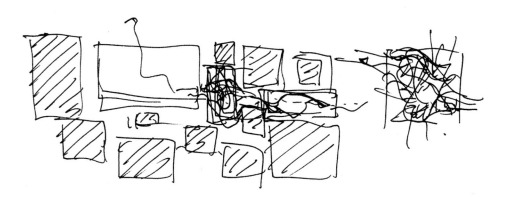

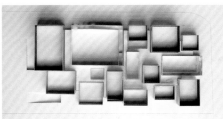

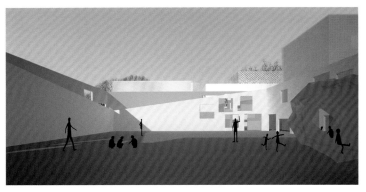
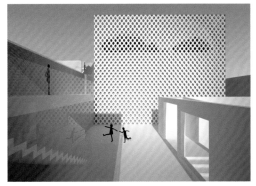

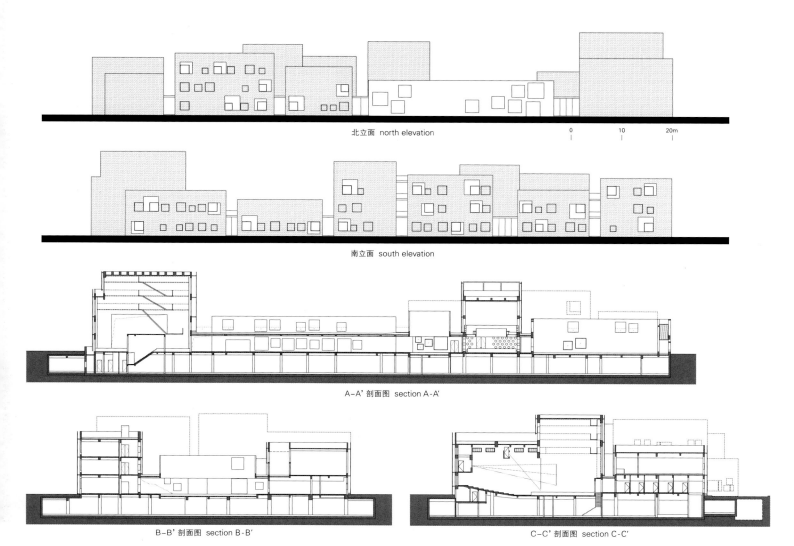

北立面 north elevation

南立面 south elevation

A-A' 剖面图 section A-A'

B-B' 剖面图 section B-B'

C-C' 剖面图 section C-C'

The building of Qingpu Youth Center decompose spaces with different functions according to their detailed characteristics of applications, turning them into relatively small-scale units and then organizes them all together through outer spaces such as courtyards, squares, streets, etc., finally making a polymer of building groups. Youth activities among spaces – the link between spaces with different functions, aimless wandering and random discovery – just as activities in a small city, are response for increasingly larger scale of the city in the process of urbanization in suburbs. We hope that under the premise that the building scale has already been enlarged, an inherent human small-scale public space could still be created and memories of scale of traditional towns could be rebuilt.

In traditional living space of regions south of the Yangtze River, external space and internal space of buildings are often treated in the same way with the former even much more important. This design obeys to such traditions while external space which is humanistic and interesting excellently corresponds to characteristics of youth personalities and activities.

A building, can also be a small city. Atelier Deshaus

项目名称：QingPu Youth Center
地点：Huake Rd., QingPu, Shanghai, China
建筑师：Liu Yichun, Chen Yifeng
设计团队：Gao Lin, Liu Qian, Wang Longhai
用地面积：10,906m²
建筑面积：14,360m²
设计时间：2009.6—2010.3
竣工时间：2012.3
摄影师：©Yao Li (courtesy of the architect) (except as noted)

©Zhang Siye(courtesy of the architect)

1 屋顶花园 2 教室 3 办公室 4 会议室 5 中空空间 6 屋顶
1. roof garden 2. classroom 3. office 4. meeting room 5. void 6. roof
三层 third floor

1 露天剧院 2 平台 3 舞蹈室 4 音乐室 5 教室 6 图书馆 7 会议室 8 办公室 9 中空空间
1. open-air theater 2. terrace 3. dancing 4. music 5. classroom 6. library 7. meeting room 8. office 9. void
二层 second floor

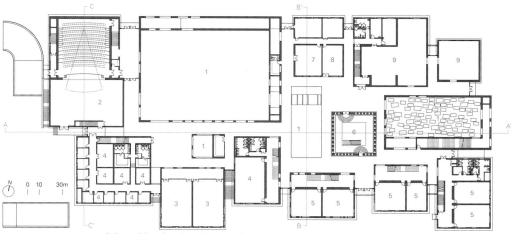

1 庭院 2 剧院 3 舞蹈室 4 音乐室 5 教室 6 图书馆 7 接待室 8 办公室 9 美术馆
1. courtyard 2. theater 3. dancing 4. music 5. classroom 6. library 7. reception 8. office 9. gallery
一层 first floor

>>86
Amann-Cánovas-Maruri
Was founded by Atxu Amann[center], Andrés Cánovas[right] and Nicolás Maruri[left] in Madrid,1987. Has focused on housing unit types and housing block reconfiguration according to contemporary urban and public space issues. Received more than seventy national and international awards. Their exhibitions were held in many countries, the recent one of which was in London.

>>144
dmvA Architecten
Established in 1997 by Tom Verschueren[left] and David Driesen[right] in Mechelen. They met each other in 1995 when they were post graduate students in Antwerpen.Both architects want to express themselves by means of architecture which is where the Dutch abbreviation dmvA(door middel ban Architectuur) stands for. Is not focused on one particular style. Act like screenwriters, turn the program into a screenplay and spaces into sequences and so allow the building to each time tell another story.

Perkins+Will
Is a global multidisciplinary architecture and design firm based on social responsibility and sustainable design established in 1935. Peter Busby is working on project across all over the world as managing director. Focuses on sustainable communities and regenerative design within numerous market sectors. As the founder and recent Chair of the Canada Green Building Council, he initiated the development of LEED in Canada. Has devoted much of his time to his profession, to the community, the development of public policy and the advancement of sustainable education and practices.

Jim Huffman, the associate principal, has experienced in urban planning and architectural design field for 25 years. When he was a design director in the Vancouver office, he was awarded a lot of projects. Well known for high-quality design with leading-edge sustainable solution.

>>156
IND
IND is an architecture and urbanism office based on Rotterdam, the Netherlands founded in 2007 by Arman Akdogan from Turkey and Felix Madrazo from Mexico. Recently won the Istanbul Technical University Campus competition and received the best architecture project award from Turkey in 2012. The office is the branch partner of the Latin American think tank Supersudaca in the Netherlands.

>>70
G.S Architects & Associates
YongMi Kim graduated from Seoul National University in 1983 and received a master's degree in 1985 from the same university. In 1990, graduated from Ecole d'Architecture Pars-Belleville and obtained her doctorate in 1991 at Universite Paris Jussieu. From 2007 to 2011, was an adjunct professor in architecture at SungKyunKwan Unversity. Has been a CEO of G.S Architects & Associates since 2004 and working as Seoul public architect since this year.

>>18
Naf Architect & Design
Tetsuya Nakazono was born in Miyazaki, Japan in 1972. Graduated from Hiroshima University, department of archtecture in 1995. Received a master's degree in 1997 and worked for Shiomi Architects and Associates. In 2001, established Naf Architect & Design. Currently is an assistant professor in Sojo University.

>>52
Ikimono Architects
Takashi Fujino was born in Gunma, Japan in 1975. Studied architecture at the Tohoku University and received a master's degree from the same university. Worked for Shimizu Coporation and Haryu Wood Studio before establishing his own studio Ikimono Architects in 2006. Currently is lecturing at Maebashi Institute of Technology.

>>128
TEN Arquitectos
Enrique Norten was born in Mexico City where he graduated from the Ibero-American University with a degree in architecture in 1978. Obtained a master of architecture from Cornell University in 1980. In 1986, founded TEN Arquitectos in Mexico City, initiating a lifelong commitment to architecture and design.

>>30
UID Architects
Was created by Keisuke Maeda who is one of the young, up and coming Japanese architects based in Hiroshima. He strongly believes that architecture should stimulate, engage and challenge the viewer. The main focus of his work is one-off tailor-made residential designs for specific clients. The success of each project is based on a set of partnerships. Since its inception in 2003, they won many prizes and participated in numerous exhibitions.

>>42
Office of Kimihiko Okada
Kimihiko Okada was born in Kanagawa, Japan in 1971. Graduated from Meiji University. Worked for Ryue Nishizawa's office for 7 years and founded his own office in 2005. Received many prizes and held a lot of exhibitions. Has taught students in several university as visiting lecturer since 2006.

>>168
Atelier Deshaus
Was establised in Shanghai in 2001. Co-founder Liu Yichun[right] and Chen Yifeng[left] received master's degree from Tongji University, department of architecture. The atelier has been involved recently in major international exhibitions on contemporary Chinese architecture in Shanghai, Beijing, and European countries including Paris, Dusseldorf, London, Barcelona, etc.. Was selected by the international Journal Architectural Record to be one of the 10 firms in year's design vanguard in 2011.

Marco Atzori
Graduated cum laude in civil engineering and architecture in Cagliari. Held a Ph.D. from University of Cagliari, with a dissertation on "The Project of the Ground". Has taught undergraduate design studios and postgraduate courses in Cagliari and Barcelona ("Intelligent Coast", Barcellona). Since 2009, he collaborates at University of Alghero. His writings have appeared in "Il Giornale dell'Architettura". Usually organizes cultural events, exhibitions and lectures about contemporary architecture. Since 2005 co-owner of architectural office Atzori+Zara with Michele Zara in Cagliari.

Silvio Carta
Is an architect and critic based in Rotterdam. Lives and works in the Netherlands, Spain and Italy, where he regularly writes reviews and critical essays about architecture and landscape for a diverse group of architecture magazines, newspapers and other medias. In 2009 he founded the Critical Agency™ | Europe.

Nelson Mota
Graduated as an architect at the University of Coimbra, in 1998. In 2006 he held the degree of Master of Architecture, Territory and Memory. He is currently a PhD researcher at the Chair of Architecture and Dwelling at Delft University of Technology. He is also active since 2004 as teaching assistant in the Department of Architecture at the University of Coimbra. As a designer he develops his work as a founding member of Comoco Arquitectos.

>>100
Weiss / Manfredi
Is at the forefront of architectural design practices that are redefining the relationships between landscape, architecture, infrastructure, and art. Marion Weiss[left] is one of the co-founder. Received her master of architecture at Yale University and her bachelor of science in architecture from the University of Virginia. Has taught design studios at Harvard University, Yale University, Cornell University, and since 1991 has been a faculty member at the University of Pennsylvania's Penn School of Design. Michael Manfredi[right] is also the co-founders. Received his master of architecture at Cornell University. Has taught design studios at Yale University, University of Pennsylvania, Princeton University, the Institute for Architecture and Urban Studies, and most recently at Harvard University.

C3: Ground Folds
All Rights Reserved. Authorized translation from the Korean-English language edition published by C3 Publishing Co., Seoul.
© 2013大连理工大学出版社
著作权合同登记06-2012年第265号

版权所有·侵权必究

图书在版编目(CIP)数据

大地的皱折：汉英对照 / 韩国C3出版公社编 ； 于风军等译. — 大连 ： 大连理工大学出版社，2013.2

书名原文：C3:Ground Folds
ISBN 978-7-5611-7649-8

Ⅰ．①大… Ⅱ．①韩… ②于… Ⅲ．①建筑学—研究—汉、英 Ⅳ．①TU

中国版本图书馆CIP数据核字(2013)第029558号

出版发行：大连理工大学出版社
　　　　　（地址：大连市软件园路80号　邮编：116023）
印　　　刷：精一印刷（深圳）有限公司
幅面尺寸：225mm×300mm
印　　张：11.75
出版时间：2013年2月第1版
印刷时间：2013年2月第1次印刷
出 版 人：金英伟
统　　筹：房　磊
责任编辑：张昕焱
封面设计：王志峰
责任校对：蒋　丽

书　　号：ISBN 978-7-5611-7649-8
定　　价：228.00元

发　行：0411-84708842
传　真：0411-84701466
E-mail：12282980@qq.com
URL: http://www.dutp.cn